TABLE OF CONTENTS

Photon, mass and gravity

Photon, mass and gravity

Oliver R Jovanovic

From the first steps (the black body radiation [1], photoelectric effect [2]…) until today (quantum, nuclear and atomic physics on the whole), there are a myriad of theoretical physics' papers with a legitimate experimental validation which, as its corner stone, have Planks constant $h = 6.626 \times 10^{-34}$ Js.

Therefore, for more than a century, thousands of physicists have worked under the assumption that h is a constant, and again, all accepted theoretical work has an experimental proof.

Hence, it is more than justified to assume that h is a number that is equal always and everywhere, but actually, it is not the case.

To summarize, it is difficult to even start contemplating the variability of an h because:

1.   All valid theories and experiments in physics lead us to $h = 6.626 \times 10^{-34}$Js.

2.   All direct measurements tell us the same.

Be that as it may, I'm inviting you to consider exactly the opposite.

In this book I will try to show you that h is a gravitational field dependent variable in the sense that h holds smaller value in region of space with stronger gravitational field and vice versa.

In my explanations I will use the Einstein's relation between the rest mass and energy ($E=mc^2$) [3], the law of energy conservation (energy is conserved always and everywhere) and the Plank's law of radiation ($E=hf$) [2] with a difference that h is a number that depends of the gravitational field.

The listed value of the constant holds for the surface of the Earth and due to the weakness of its gravitational field, changes in its value with the height are small; nonetheless the change exists. It is impossible to note a change in the direct measurement (all instruments and its parts differ with the altitude), but it is possible to

asses it indirectly (Pound RV, Rebka GA (1959) [4] if we expect it and if we interpret the results correctly.

All in all, I will try to justify the validity of the new view for the gravitational red and blue shift, through a correct interpretation [5] of the "Harvard tower experiment" [4] with a help from the stated relations. In the last part of this book I will present some of the numerous consequences and I will speculatively consider the new physical constructions enabled by this approach.

Energy and the rest mass

Most famous equation [3] states a correct link between the rest mass and energy, as author writes "mass is energy and energy is the mass".

In the following few examples, I will illustrate this relationship in the way that professors around the world explain it, and, in the end, I will add my own example for the mass increase in the gravitational field.

Standard examples:

1. Work against the nuclear forces

I0394110

Nuclear fusion (energy of the stars), when fusing lighter elements into heavier, the overall mass of the initial atoms decreases.

Example 1: "proton-proton chain reaction [6]"-four protons (ionized hydrogen atoms) create helium nucleus and as a result of partial mass loss; the energy equivalent is released. In line with Sir Arthur Stanley Eddington (1882 –1944) [8], Hans Albrecht Bethe (1906–2005) [9] and others, approximate over all mass change can be estimated simply as: $\Delta m \approx 4$(mass of proton) -mass of helium nucleus [10]

$\Delta m \approx 4(1.672 \times 10^{-27}$ kg$) - 6.645 \times 10^{-27}$ kg $\approx 0.043 \times 10^{-27}$ kg - mass defect

Associated with mass change[1] is the energy release[2];this difference of initial energy of constituents is assessed by stated relation:

$\Delta E = \Delta mc^2 \approx 0.043 \times 10^{-27}$ kg $(3 \times 10^8 \frac{m}{s})^2 \approx 3.87 \times 10^{-12}$ J

or in eV (electronvolts)

---

[1]Mass defect
[2]Bond energy; the overall energy is always conserved, but bulk of bond energy is radiated away from system.

$\Delta E \approx 3.87 \times 10^{-12}$ J$\approx 3.87 \times 10^{-12} \times 6.242 \times 10^{18}$ eV$\approx 24\ 000\ 000$ eV≈ 24 MeV per one cycle.

Among the other things, this means that if we take helium nucleus in one hand then, with the other, we pluck nucleons, one by one:

a. We have to work against nuclear forces
b. We have to invest energy equal to that work
c. In doing so, we will increase the mass of each constituent

2. Work against the elastic forces

One of the standard examples for relation between the mass and energy is about an elastic spring (one variant is in the book "Einstein for dummies [7]" page 139-Serbian edition).

The unstressed elastic spring has the smallest mass, but if we stretch or squeeze it, we'll increase its mass in the known equivalence.

Example 2: Calculate the rise of a potential energy of the elastic spring and associated mass gain, if we stretch it by 0.5m. Spring constant is k=400 N/m [11]

(F=kx – force exerted on spring, equal and opposite to the spring elastic force. x-displacement)

Answer:

$$\Delta E = \text{work against elastic force} = \int_0^x F dx = \int_0^x kx dx = \frac{kx^2}{2} = \frac{400\frac{N}{m}(0.5m)^2}{2} = 50 \text{ J}$$

Additional potential energy is responsible for the rest mass increase as:
$\Delta E = \Delta m c^2$; or a mass increase can be estimated as:

$$\Delta m = \frac{\Delta E}{c^2} \approx \frac{50 \text{ J}}{(3*10^8\frac{m}{s})^2} \approx 5.56 * 10^{-16} kg$$

This kind of spring can have around m=0.1 kg of its original mass, and even though the increase of mass is difficult to measure, nonetheless it exists.

m=0.1 kg   - mass of the unstressed spring
m'=m+Δm $\approx$ 0.1 kg + 5.56x10^{-16} kg   - mass of the stressed spring

Therefore, here also stands, when we want to stretch or squeeze the elastic spring:
a. We have to work against the elastic forces

4

b.   We have to invest the energy which is equal to that work

c.   In doing so, we will increase the mass of the spring

3.   Work against the magnetic forces

At the after-graduate exam party of my distinguished colleague, Aleksandar Bogicevic, one of the professors asked us: "What arrangement has a bigger mass; the two magnets threaded onto a piece of wood so that they are repelled (separated by its own magnetic forces) or the same two magnets placed reversed-close together and bonded by its own magnetic forces?"

(Similar example is in the book "Einstein for dummies [7]" page 140-Serbian edition)

The answer is, of course, two magnets that are in repulsive configuration because when we want to separate them from attractive configuration:

a. We have to work against magnetic forces

b. We invest the energy equal to that work,

c. In doing so, we increase the mass of magnets.

4.   Work against the gravitational force (my personal contribution)

If we take a cat and raise it on a table top from the floor:

a. We have to work against the gravitational force of the Earth

b. We invest the energy equal to that work,

c. In doing so we increase the mass of the cat.

Elaboration:

Is there a gravitational force? Yes.

Newton's[3] equations for the gravitational force and his second law are precise enough for story in this chapter.

Newton's equation for gravitational force $F = G\frac{m_1 m_2}{R^2}$

Where $G \approx 6.67 \times 10^{-11} \frac{Nm^2}{kg^2}$ is gravitational constant[4],m_1 and m_2 are masses of objects 1 and 2, respectively, and R is the distance between the centers of them.

Newton's equation for second law: $F = ma$ (the force exerted on object with mass m will result in objects' acceleration a.

Every loose object close to Earth's surface, regardless of its own mass, has the acceleration of approximately $g \approx 9.81 \frac{m}{s^2}$. Why is that a case?

Let's say that object has a mass of m_1, at the Earth surface, any loose object will start to move toward the center of the planet with acceleration of $g \approx 9.81 \frac{m}{s^2}$. This can be assessed with some simple demonstrations. Then we say that there have to be a force that causes this behavior or $F=m_1 g$ (according to the Second Newton's law). On the other hand, according to Newton's law of gravity, the same force can be represented as $F = G\frac{m_1 m_2}{R^2}$, and because F=F, $m_1 g = G\frac{m_1 m_2}{R^2}$, now we see that there is a nonzero m_1 present at both sides of equation. Therefore, we can reduce equation to $g = G\frac{m_2}{R^2}$, and here we see that the acceleration of any body at the Earth's surface does not depend of its own mass m_1 (every object has same acceleration g), but it is dependent of the Earth's mass m_2.

Note that "Cavendish experiment 1797–1798"was the first correct assessment for gravitational attraction coefficient, and it was the first effective procedure that measured the mass of the Earth as $m_2 = \frac{gR^2}{G}$ and an average density as $\rho = \frac{m_2}{V} = \frac{3g}{4\pi RG}$, g and R- Earth radius, which was previously determined by Newton, Eratosthenes (276-194 BC) and others.

So, there is a gravitational pull on objects at the Earth's surface and its value is F=mg, where F is a gravitational force exerted on mass m by the abovementioned planet, m is a mass of the object and $g \approx 9.81 \frac{m}{s^2}$ the acceleration of the loose object.

---

[3] Sir Isaac Newton (December 1642 –March 1726)
[4] First time measured by Henry Cavendish (October 1731 – February 1810),one of the most important experiments in the history of science, "the Cavendish experiment1797–1798"

If an object rests on the ground, there is a gravitational force exerted on it, but there is no motion and therefore no acceleration, and in this case, the gravitational force is balanced by the reactive force of the ground.

Let's get back to a cat.

Example 3: Calculate the rise in a potential energy of the cat and the associated mass of the cat increases, when we raise it from the floor to a table top, the desk is H=1m high and the mass of the cat at the floor is m=5kg.

Answer:

ΔE=work against the gravitational force=$\int_0^H F dx = \int_0^H mg dx = mgH$ =5kg 9.81 $\frac{m}{s^2}$ 1m $\approx$ 49.05 J

(Again, this kind of calculation is precise enough because g is small, and difference in g and m for 1meter at the Earth surface is small).

The additional potential energy is responsible for rest mass increase as:
$\Delta E = \Delta mc^2$; or mass increase can be estimated as:

$\Delta m = \frac{\Delta E}{c^2} \approx \frac{49.05\ J}{(3*10^8\frac{m}{s})^2} \approx 5.45 * 10^{-16} kg$

m=5 kg   - the mass of the cat on the floor
m'=m+$\Delta m \approx$ 5 kg + 5.45x10^{-16} kg   - the mass of the cat on the table

It is my personal opinion that these small changes in the rest mass are just a glimpse on what is really happening in extreme conditions, and therefore:

In other part of this book I will try to show you that there are places in this Universe where mass of the named cat will change in great amount.
In other words, I will try to prove that the mass of the cat can change 2 and more times at the same distance. Note that the same applies for the particles, so I will say that the mass of the quark, electrons, protons, neutrons and specific atoms is estimated for our environment and that there are locations in this Universe where the mass of the abovementioned particles is significantly different.

The vast majority of physicists will agree on a principle with the statements given under numbers 1, 2, and 3, but the claims and consequences expressed in 4 are generally hard to accept. And this is just the beginning.

Energy conservation in a free fall

If we accept it as a correct one, the doctrine that is being taught in all primary and secondary schools as well as in colleges and universities; within classical mechanics we teach the connection between the work in the gravitational field, potential and kinetic energy through energy conservation laws. In other words, the left and the right shoe have equal energy if they are at the same height, but if we raise one of them to a higher level, we have to work against gravitational force and, in doing so, we increase the energy (potential energy) of the elevated shoe. The increase of the potential energy is identical to the work performed.

Now, the shoe at a higher altitude has more energy than the one on the floor (it holds more of potential energy and therefore more of total energy).

If we now let the raised shoe fall towards the baseline level, its potential energy will go into a kinetic one, but the total energy (the sum of the potential and the kinetic one) will remain constant at all times.

In a split of a second, just before flying shoe hits the ground, we observe shoes next to each other and we conclude that the falling shoe moves at expected speed and that its energy remains greater than one that did not move.

An external work exerted to hoist a shoe has increased its potential energy, which has converted completely into kinetic just before the impact; so we worked to increase the energy and this work (the energy) did not vanish in the free fall; that is: the energy is fully maintained throughout the drop.

In physics we should learn that object retains its energy during the free fall, in a way that the potential energy continuously converts into kinetic energy, close to Earth's surface holds: $mg H = mv^2/2$. In other words, even though the body experience increases in speed, its energy is constant. (m-mass of the body, H –fall distance, v-the speed of the body relative to a starting point, g=9.81 m/s^2- gravitational acceleration at the Earth's surface).

My personal contribution [5]: we will extend the law of energy conservation for the free-falling photon (the free fall of the photon-photon moves towards the source of the gravitational force); my personal opinion is that the photon keeps its energy in the free fall even though its frequency increases. From which it have to be concluded that the value of the Plank's constant is different at different heights.

$E=E' \Rightarrow hf=h'f'$

E – Photon energy at some elevation

E' – Photon energy after some displacement towards the gravitational source (after some time in a free fall)

Photon falling into a gravitational field exhibits a blue shift [4]; the frequency of the photon increases $f'>f$ which directly implies $h' < h$ reduction for values of Plank's constant, because photons energy is constant in the free fall $E = E'$ regardless of its change of frequency.

In other words, the Plank's constant h holds smaller values as we get closer to the massive bodies; or the Plank's constant is a variable, in a way that, it is smaller at stronger gravitational fields and vice versa.

This change is expressed by equation $h' = \dfrac{hf}{f'}$

(h' Planck's constant at lower altitude, f'- photon frequency at lower altitude; h Planck's constant at higher altitude, f – photon frequency at higher altitude)

Now would be a right moment for a reader to review my short communication published with "Journal of Applied & Computational Mathematics" under a title: Correct Interpretation for "Harvard Tower Experiment" or Law of Energy Conservation for "Free Fall" Photon [5].

In short, my personal most important claims [5]

1.   If we elevate the body in the gravitational field, we will increase its mass for the expected rate (at Earth surface) $\Delta m = \frac{E}{c^2} = \frac{mgH}{c^2}$

2.   In its movement toward or from the source of the gravitational force, the photon will experience the change in frequency, but its energy will remain constant. This directly implies that the Plank's constant is a field dependent variable $h'=\frac{hf}{f'}$

Future physics?

If we begin to observe a nature thought physics with adjustment noted in the previous subtitle, it will lead us to some new and interesting domains, some of which are unthinkable under the current state of affairs.

Some of the consequences are straightforward and easy to notice, some are hardly noticeable, but only possible with help from stated assumptions, and another group of hypotheses only has some philosophically-speculative links to this point of view.

1.   Yellow photon which starts its voyage at the Earth's surface is more red near the surface of the Moon[5], and is more blue[6] when it arrives at the surface of the Sun. It has a constant energy the entire trip regardless of its change in color (i.e. in this case energy is constant regardless of its change in frequency).

2.   The masses of quark, electron, proton, neutron, specific atom, molecule and every object depends of gravitational field. If we lift named objects-we increase their masses, and if we lower them down-we decrease their masses.

3.   With a change in mass it is logical to assume a change in size (length, height, width) of the named objects. In other words, cat, electron, quark, atom… are

---

[5] gravitational red shift
[6] gravitational blue shift

smaller under a table than they are on the table and they are even bigger if we raise them to the ceiling.

4. Since the speed of light c^7 is a constant everywhere, it is interesting to note that the increase of a photon frequency in a free fall causes its wavelength to decrease; the photon becomes smaller in a free fall, (some similarity with the situation under number 3).

5. Gravitational force and acceleration equation (my personal work; it is the law which I'm stating). One of the major consequences of the rest mass reduction in the gravitational field are new equations for the gravitational force and acceleration, where originally, the equation is derived from the free fall of the small object towards a much bigger one; the generalized equation form is a necessity introduced to accommodate the Newton's third law^8 and proper distance for maximum energy density(see number 6 a).

The gravitational force F between bodies 1 and 2 at the distance R, $G \approx 6.67 \times 10^{-11} \frac{Nm^2}{kg^2}$, C- the speed of light, m_1 and m_2 the initial masses of the bodies 1 and 2 (as energy measurement at initial conditions m_1 and m_2 does not change during the motion if the system is isolated; on the other hand, the rest mass changes even in isolated systems, see number 6 b).

$$F = G \frac{m_1 m_2}{R^2} \frac{1}{1 - \frac{2G(m_1 + m_2)}{RC^2}}$$

And due to $F=F_{12}=F_{21}=m_1 a_1$, the gravitational acceleration a; of the body 1 toward body 2 is derived from $m_1 a = G \frac{m_1 m_2}{R^2} \frac{1}{1 - \frac{2G(m_1+m_2)}{RC^2}}$ as:

$$a = G \frac{m_2}{R^2} \frac{1}{1 - \frac{2G(m_1 + m_2)}{RC^2}}$$

Functions F and a are defined for:

---

^{7} c=λf (speed of light=wavelength×frequency)
^{8} "When one body exerts a force on a second body, the second body simultaneously exerts a force equal in magnitude and opposite in direction on the first body."

$$\frac{2G(m_1+m_2)}{c^2} < R$$ (In this region the gravitational force is attractive with large values as R goes near$\frac{2G(m_1+m_2)}{c^2}$)

$$0 < R < \frac{2G(m_1+m_2)}{c^2}$$ (In this region the gravitational force is repulsive with large values as R goes near$\frac{2G(m_1+m_2)}{c^2}$).

6 a.   A realistic limit in energy density[9]. The gravitational force changes its nature near R→ $\frac{2G(m_1+m_2)}{c^2}$ from extremely attractive, through undefined, to extremely repulsive. As we are going to see in the next few examples, there is no force in nature that is stronger than the gravitational force in these regions, these conditions impose realistic limits in mass (energy) densities. At regions near $R \to \frac{2G(m_1+m_2)}{c^2}$ and $0 < R < \frac{2G(m_1+m_2)}{c^2}$, the gravitational force is repulsive and much stronger than any other force in the Universe so nothing can squeeze matter much more than a certain maximum density. The radius of that region is estimated as$R \approx \frac{2GM}{c^2}$, where M is a mass encircled by the sphere of a radius R. Or the maximum energy density can be estimated from $R \approx \frac{2G}{c^2}\frac{E}{c^2} = \frac{2GE}{c^4}$.

Whatever reaches mass density or energy densities, and is larger than this; explodes violently. In other words, slightly denser objects in this Universe can exist, but objects move apart under predominant repulsive gravitational force; in addition to that, everything that happens in this Universe will obey laws of energy, momentum and angular momentum conservation.

Example 4: a) How much mass and energy you can squeeze into sphere with radius of 1 m, before gravitational force becomes repulsive.

$$R \approx \frac{2GM}{c^2} \text{Or } M \approx \frac{RC^2}{2G}$$

$$M \approx \frac{1m \times (3\times10^8\frac{m}{s})^2}{2\times6.67\times10^{-11}\frac{Nm^2}{kg^2}} \approx 6.75 \times 10^{26} kg$$ This is approximately amount of mass you can

put in region encircled by 1m; any more mass and gravitational force becomes repulsive.

---

[9] This is the primary cause for what appears to be Universe accelerated expansion i.e. real reason behind "dark energy".

$$V = \frac{4}{3} \times R^3 \times \pi \approx \frac{4}{3} \times (1m)^3 \times 3.14 \approx 4.19 \ m^3$$

$$\rho = \frac{M}{V} \approx \frac{6.75 \times 10^{26} kg}{4.19 \ m^3} \approx 1.61 \times 10^{26} \frac{kg}{m^3}$$

Associated energy density η:

$$\eta = \frac{E}{V} = \frac{mc^2}{V} = \rho c^2 \approx 1.61 \times 10^{26} \frac{kg}{m^3} (3 \times 10^8 \frac{m}{s})^2 \approx 1.5 \times 10^{43} \frac{J}{m^3}$$ This is

approximately the amount of energy you can put in region encircled by 1m; any additional energy and gravitational force becomes repulsive.

   b) How much mass and energy you can squeeze into sphere with radius of 10^{10} m, before gravitational force becomes repulsive.

$$M \approx \frac{10^{10}m \times (3 \times 10^8 \frac{m}{s})^2}{2 \times 6.67 \times 10^{-11} \frac{Nm^2}{kg^2}} \approx 6.75 \times 10^{36} kg$$ This is approximately amount of mass you can

put in region encircled by 10^{10}m; any additional mass and gravitational force become repulsive.

$$V = \frac{4}{3} \times R^3 \times \pi \approx \frac{4}{3} \times (10^{10}m)^3 \times 3.14 \approx 4.19 \times 10^{30} m^3$$

$$\rho = \frac{M}{V} \approx \frac{6.75 \times 10^{36} kg}{4.19 \times 10^{30} m^3} \approx 1.61 \times 10^6 \frac{kg}{m^3}$$

$$\eta = \frac{E}{V} = \frac{mc^2}{V} = \rho c^2 \approx 1.61 \times 10^6 \frac{kg}{m^3} (3 \times 10^8 \frac{m}{s})^2 \approx 1.5 \times 10^{23} \frac{J}{m^3}$$ This is

approximately the amount of energy you can put in region encircled by 10^{10}m; any additional energy and gravitational force become repulsive.

   c) How much mass and energy you can squeeze into sphere with radius of 10^{21} m (the distance comparable with the radius of the Milky Way galaxy), before the gravitational force becomes repulsive.

$$M \approx \frac{10^{21}m \times (3 \times 10^8 \frac{m}{s})^2}{2 \times 6.67 \times 10^{-11} \frac{Nm^2}{kg^2}} \approx 6.75 \times 10^{47} kg$$

$$V = \frac{4}{3} \times R^3 \times \pi \approx \frac{4}{3} \times (10^{21}m)^3 \times 3.14 \approx 4.19 \times 10^{63} m^3$$

$$\rho = \frac{M}{V} \approx \frac{6.75 \times 10^{47} kg}{4.19 \times 10^{63} m^3} \approx 1.61 \times 10^{-16} \frac{kg}{m^3}$$

$$\eta = \frac{E}{V} = \frac{mc^2}{V} = \rho c^2 \approx 1.61 \times 10^{-16} \frac{kg}{m^3} (3 \times 10^8 \frac{m}{s})^2 \approx 15 \frac{J}{m^3}$$

d) How much mass and energy you can squeeze into sphere with radius of 10^{23} m (the distance comparable with a gap between some galaxies), before the gravitational force becomes repulsive.

$M \approx \dfrac{10^{23}m \times (3 \times 10^8 \frac{m}{s})^2}{2 \times 6.67 \times 10^{-11} \frac{Nm^2}{kg^2}} \approx 6.75 \times 10^{49} kg$ This is approximately the amount of mass you can put in region encircled by 10^{23}m; any additional mass and gravitational force become repulsive.

$$V = \frac{4}{3} \times R^3 \times \pi \approx \frac{4}{3} \times (10^{23}m)^3 \times 3.14 \approx 4.19 \times 10^{69} m^3$$

$$\rho = \frac{M}{V} \approx \frac{6.75 \times 10^{49} kg}{4.19 \times 10^{69} m^3} \approx 1.61 \times 10^{-20} \frac{kg}{m^3}$$

$\eta = \dfrac{E}{V} = \dfrac{mc^2}{V} = \rho c^2 \approx 1.61 \times 10^{-20} \frac{kg}{m^3} (3 \times 10^8 \frac{m}{s})^2 \approx 0.0015 \frac{J}{m^3}$ This is approximately the amount of energy you can put in region encircled by 10^{23}m; any additional energy and gravitational force become repulsive.

In these previous few examples we saw that mass (energy) density of some region is something that defines the nature of the gravitational force; in addition to that, a critical density is size dependent, so we can have the same two objects in this Universe that accelerate toward or away from each other just because they find themselves to be in a place with attractive or repulsive surrounding configurations.

Example[10] 5.a. ("dark energy"):Calculate an average density of the large region of, seemingly empty space, encircled by 10^{23}m if you observed two galaxies, located on the opposite sides of the region, that accelerate from each other with the acceleration of $10 \frac{m}{s^2}$ (the relative speed of galaxies is small).

Answer: It is a custom to present the attractive forces and acceleration as negative values; in this example I did it vice versa due to a maximum simplification of the stories. In this example, the acceleration is attractive when it is positive, and therefore, it shows itself to be negative in repulsive circumstances.

---

[10] Calculations in example 5 are demonstrative and educational, but somewhat imprecise. Hopefully real-life calculations will use multitude of data and supercomputers.

The acceleration of the objects at the surface of the region:

$$a = G\frac{M}{R^2}\frac{1}{1 - \frac{2GM}{RC^2}}$$

R-radius of the region, M –mass of the region, M>>m-mass of the galaxy, etc.

$a = -5\frac{m}{s^2}$-The acceleration of the galaxy away from the center of the region (a repulsive gravitational force).

$$-5\frac{m}{s^2} = 6.67 \times 10^{-11}\frac{Nm^2}{kg^2}\frac{M}{(10^{23}m)^2}\frac{1}{1 - \frac{2\times6.67\times10^{-11}\frac{Nm^2\times M}{kg^2}}{10^{23}m\times(3\times10^8\frac{m}{s})^2}}$$

Solution for M
$M \approx 6.7466 \times 10^{46}kg$ Mass of the named region
Volume

$$V = \frac{4}{3} \times R^3 \times \pi \approx \frac{4}{3} \times (10^{23}m)^3 \times 3.14 \approx 4.19 \times 10^{69}m^3$$

Density of the region

$$\rho = \frac{M}{V} \approx \frac{6.75 \times 10^{49}kg}{4.19 \times 10^{69}m^3} \approx 1.61 \times 10^{-20}\frac{kg}{m^3}$$

Example 5.b. ("dark matter"). Use density from example 5.a. to calculate the mass of the unknown material which occupies sphere with center at the center of the Sun and with a radius up to us here.
Answer: $R \approx 1.5 \times 10^{11}m$ the distance to the Sun

$$m = \rho V = \rho \times \frac{4}{3} \times R^3 \times \pi$$

$$\approx 1.61 \times 10^{-20}\frac{kg}{m^3} \times \frac{4}{3} \times (1.5 \times 10^{11})m^3 \times 3.14 \approx 2.26 \times 10^{14}kg$$

Which is approximately: $\frac{2.26\times10^{14}kg}{5.75\times10^{13}kg} \approx 4$ Mount Everest of probably "wimp" mater scattered around the Sun, up to us.

15

Again, these are some demonstrative calculations for what scientists call "the dark energy" and "the dark matter", hopefully much better results will be obtained with more precise data and calculations.

6 b. The rest mass and the invariant mass[11].

If you take an apple from the ground and raise it to the table top; you will increase its mass, and if you take the same apple from the table top and put it back to the ground; you will decrease its mass. In the first case you will invest energy and you will increase its mass, in other case, the apple will give same energy back to you. And if you could have very precise scale attached to your arm you could measure this change in mass continuously. Hopefully, I succeeded in assuring you in previous statements by this page.

On the other hand, if the apple drops from the table top, its original invariant mass will not change, for sure its rest mass will decrease continuously and its speed will rise[12] as a result of that; but thing that creates and exerts the gravitational pool will remain constant in a free fall. Original rest mass on the table top is a measurement of all the energies in its free motion to the ground.

Example 6: a) Asses the rest mass reduction of the apple that falls from the table after H=0.3m, the original rest mass of the apple is m=0.5 kg.

           b)   Write equation for the force and acceleration in the named process.

Answer: a) (Approximation for weak fields) $mc^2=m'c^2+m'gH$

Where m-original rest mass, apple on the table top; m'-reduced rest mass, after fall of H=0.3m; $g=9.81\frac{m}{s^2}$ the gravitational acceleration at the Earth surface.

$$mc^2-m'c^2=m'gH$$
$$(m-m')c^2=m'gH$$
$$\Delta mc^2=m'gH \quad \text{(at Earth's surface } m\approx m')$$
$$\Delta mc^2=mgH$$

---

[11] All masses m_1, m_2 and M mentioned in 6 a.
[12] In extremely strong gravitational fields the rest mass can go to zero as speed of the falling object goes up to the speed of light.

$$\Delta m \approx \frac{mgH}{c^2} = \frac{0.5kg9.81\frac{m}{s^2}0.5m}{(3\times10^8\frac{m}{s})^2} \approx 1.635 \times 10^{-17}kg$$ The rest mass loss after fall of 0.3 m

b) The equation for the force exerted on the apple and its acceleration in this process.

$F = G\frac{mm_2}{R^2}\frac{1}{1-\frac{2G(m+m_2)}{RC^2}}$ The force exerted on the apple, m-original rest mass of the apple, m_2-mass of the Earth etc.

$a = G\frac{m_2}{R^2}\frac{1}{1-\frac{2G(m+m_2)}{RC^2}}$ The acceleration of the apple in the same process.

Note that in the expression of the force and acceleration figures the original rest mass of the apple. In other words, we use invariant mass in assessments of gravitational pull and gravitational acceleration.

The Earth, along with the Moon, moves around the Sun in elliptical orbit, sometimes it is closer and at other times is further from the Sun, but the variation of gravitational pull between the Earth and the Moon due to the reduction of their rest masses is not noticed, therefore, the motion energy of the Earth and Moon as well as their rest masses exert and feel gravitational pull.

Loosely related comment: Photon generates its own gravitational field; and it is attracted by gravitational fields created by other energies (masses).

Example 7: Calculate rise in the potential energy of the cat, from the example 3, and associated mass gain, when we raise it from the surface of an average neutron star [12] to the 1m above it. For the mass of the average neutron star takes 1.2 solar masses [13], and let's say that it has radius of 13.5 km.
Answer:
Mass m=1.2m_{sun}≈1.2*1.99x10^{30}kg≈2.39x10^{30} kg
Radius R=13.5 km=13500m=1.35x10^4 m
Let us first calculate the gravitational acceleration caused by neutron star at the surface g', and one meter above that g".

$$g'= 6.67 \times 10^{-11}\frac{Nm^2}{kg^2}\frac{2.39 \times 10^{30}\text{ kg}}{(1.35 \times 10^4\text{ m})^2}\frac{1}{1-\frac{2 \times 6.67 \times 10^{-11}\frac{Nm^2}{kg^2} \times (5kg+2.39 \times 10^{30}\text{ kg})}{1.35 \times 10^4\text{ m} \times \left(3 \times 10^8\frac{m}{s}\right)^2}} \approx 1.18588 \times 10^{12}\frac{m}{s^2}$$

$$g''= 6.67 \times 10^{-11}\frac{Nm^2}{kg^2}\frac{2.39 \times 10^{30}\text{ kg}}{(1.35 \times 10^4+1m)^2}\frac{1}{1-\frac{2 \times 6.67 \times 10^{-11}\frac{Nm^2}{kg^2} \times (5kg+2.39 \times 10^{30}\text{ kg})}{(1.35 \times 10^4\text{ m}+1m) \times \left(3 \times 10^8\frac{m}{s}\right)^2}} \approx 1.18567 \times 10^{12}\frac{m}{s^2}$$

For our calculation here, it is good enough to consider that g'= g" $\approx 1.186 \times 10^{12}\frac{m}{s^2}$
Therefore we can use the equation:

ΔE=work against the gravitational force$= \int_0^H F dx = \int_0^H mg'dx = mg'H$ =5kg

$1.186 \times 10^{12}\frac{m}{s^2}1m \approx 5.93 \times 10^{12}$ J

The additional potential energy is responsible for the rest mass increase as:

$\Delta E = \Delta m c^2$; or mass increase can be estimated as:

$\Delta m = \frac{\Delta E}{c2} \approx \frac{5.93 \times 10^{12}\text{ J}}{(3 \times 10^8\frac{m}{s})^2} \approx 0.0000659\ kg \approx 6.59 \times 10^{-5}kg$

m=5 kg   - the cat's mass at the surface of the neutron star
m'=m+Δm $\approx$ 5 kg + 6.59 $\times$ 10^{-5}kg   - the cat's mass one meter above the named
surface

A change in the cat's mass for distance of 1 m at exampled neutron star is
$\frac{6.59*10^{-5}kg}{5.45\ x10^{-16}kg} \approx 1.21 \times 10^{11}$=121000000000 times greater than the change for same distance
at the surface of the Earth; nonetheless the change is a small relative to its original mass.
Let us now consider even bigger object, but first we will estimate a maximum possible
radius and a mass for neutron star-like object.

Example 8: Estimate a maximum possible radius and a mass for neutron star-like
object. For an average density use the data from previous example: $\rho = \frac{M}{V} \approx$

$\frac{2.39 \times 10^{30}\text{ kg}}{\frac{4}{3} \times 3.14 \times (1.35 \times 10^4\text{ m})^3} \approx 2.32 \times 10^{17}\frac{kg}{m^3}$.

Answer: In the region near1 $\approx \frac{2GM}{RC^2}$, the gravitational force changes from
extremely attractive; through undefined to extremely repulsive, therefore an object like
neutron star will explode.

Calculation for radius:

$1 \approx \frac{2GM}{RC^2}$ (Multiply both sides with $\frac{1}{R^2}$)

$$\frac{1}{R^2} \approx \frac{2GM}{R^3C^2} \approx \frac{2G\rho}{C^2}$$

$$R \approx \frac{C}{\sqrt{2G\rho}} \approx \frac{3 * 10^8 \frac{m}{s}}{\sqrt{2 \times 6.67 \times 10^{-11} \frac{Nm^2}{kg^2} \times 2.32 \times 10^{17} \frac{kg}{m^3}}} \approx 53926 \, m$$

As calculations show radius of neutron star-like object cannot exceed around $54000 \, m$.

Calculation for mass for this type of object:

$$M = \rho V \approx \rho \frac{4}{3} R^3 \pi \approx 2.32 \times 10^{17} \frac{kg}{m^3} \frac{4}{3} (53926m)^3 3.14 \approx 1.52 \times 10^{32} kg$$

Or around $\frac{M}{M_{sun}} \approx \frac{1.52 \times 10^{32} kg}{1.99 \times 10^{30} kg} \approx 76$ solar masses[13]

Note that denser objects in this Universe are possible (see page 11;$\rho_{max} \approx 6.7466 \times 10^{26} \frac{kg}{m^3}$), but calculation suggests that those denser objects can't be similar to neutron star. Again, for neutron star-like objects the maximal mass is in order of $10^{32} kg$ and the maximal radius is at best in order of $5 \times 10^4 m$. There might be some other mechanisms that will cause real life neutron star explosion much sooner, but neutron star-like object definitely can't exist beyond the calculated values.

Example 9:Observe a special kind of 5kg cat on the surface of the neutron star-like object that has a mass of $m = 1.77808 \times 10^{31} kg$, and a density of $2.32 \times 10^{17} \frac{kg}{m^3}$. What would be a rest mass increase if we lift it one meter above the mentioned surface?

Answer:

Radius of the star: $R = (\frac{m}{\frac{4}{3}\pi\rho})^{\frac{1}{3}} \approx (\frac{1.77808 \times 10^{31} kg}{\frac{4}{3} \times 3.14 \times 2.32 \times 10^{17} \frac{kg}{m^3}})^{\frac{1}{3}} \approx 26355.11865$ m.

---

[13] Radius and density of the neutron stars are not known precise enough which can cause this estimate to be off bay order of magnitude. Calculated values for the mass and radius are imprecise even with used numbers, nonetheless one can achieve greater accuracy iteratively, but that would excide this book format.

Gravitational acceleration for the cat at the surface:

$$g' = 6.67 \times 10^{-11} \frac{Nm^2}{kg^2} \frac{1.77808 \times 10^{31} kg}{(26355.11865m)^2} \cdot \frac{1}{1 - \dfrac{2 \times 6.67 \times 10^{-11} \frac{Nm^2}{kg^2} \times (5kg + 1.77808 \times 10^{31} kg)}{26355.11865m \times \left(3 \times 10^8 \frac{m}{s}\right)^2}} \approx 2.068 \times 10^{18} \frac{m}{s^2}$$

Gravitational acceleration for the cat, one meter above the surface:

$$g'' = 6.67 \times 10^{-11} \frac{Nm^2}{kg^2} \frac{1.77808 \times 10^{31} kg}{(26355.11865m + 1m)^2} \cdot \frac{1}{1 - \dfrac{2 \times 6.67 \times 10^{-11} \frac{Nm^2}{kg^2} \times (5kg + 1.77808 \times 10^{31} kg)}{(26355.11865m + 1m) \times \left(3 \times 10^8 \frac{m}{s}\right)^2}} \approx 4.404 \times 10^{16} \frac{m}{s^2}$$

In the first assessment of the mass increase we will use smaller g, $g'' \approx 4.404 \times 10^{16} \frac{m}{s^2}$

ΔE=work against the gravitational force=$\int_0^H F dx = \int_0^H mg'' dx = mg'' H =$
$5kg \times 4.404 \times 10^{16} \frac{m}{s^2} \times 1m \approx 2.202 \times 10^{17} J$

A mass increase can be estimated as, at very least, to be:
$\Delta m = \frac{\Delta E}{c^2} \approx \frac{2.202 \times 10^{17} J}{(3*10^8 \frac{m}{s})^2} \approx 2.45 kg$

This is a very rough estimate because, at these conditions m and g differ with height significantly, there is a major error in previous calculation about the mass increase; luckily, we can calculate as precise as we want with a help of an old Archimedes trick for assessment of number π.

I will calculate the mass increments at, let's say, 4 different elevations:
The first from 0 m to 0.25m, the second from 0.25m to 0.5m and the third from 0.75m to 1 m above the star; for every increment I will use gravitational acceleration at highest point and the rest mass at its lowest point, in that way I will have more precise calculation and I will be sure that the rest mass increase must be greater than the one we got in calculations. The more different point you make, the more precise your calculations will be.

Here $\Delta m \approx \Delta m_1 + \Delta m_2 + \Delta m_3 + \Delta m_4$
Calculations similar with previous and under a recently stated guidelines yields:

$$\Delta m_1 \approx \frac{m_0 g_1 h_1}{c^2} \approx \frac{5kg \times 1.66 \times 10^{17}\frac{m}{s^2} \times 0.25m}{(3 \times 10^8 \frac{m}{s})^2} \approx 2.30kg; \text{ the mass increase for first}$$

0.25m above star, m_0-mass at the surface, g_1-gravitational acceleration at 0.25 m above star surface, h_1 – rise distance from the star surface to 0.25 m above.

$$\Delta m_2 \approx \frac{m_1 g_2 h_2}{c^2} \approx \frac{(5kg + 2.3kg) \times 8.62 \times 10^{16}\frac{m}{s^2} \times 0.25m}{(3 \times 10^8 \frac{m}{s})^2} \approx 1.75 \ kg; \text{ the mass increase for}$$

the second 0.25m above star, $m_1 \approx 7.3$ kg-mass at the 0.25m above surface, g_2- gravitational acceleration at 0.5 m above star surface, h_2 – rise distance from 0.25 m to 0.5m above stars surface.

$$\Delta m_3 \approx \frac{m_2 g_3 h_3}{c^2} \approx \frac{(7.3kg + 1.75kg) \times 5.83 \times 10^{16}\frac{m}{s^2} \times 0.25m}{(3 \times 10^8 \frac{m}{s})^2} \approx 1.47kg; \text{ the mass increase for}$$

the third rise of 0.25m above star, $m_2 \approx 9.05$ kg-mass at the 0.5m above surface, g_3- gravitational acceleration at 0.75 m above star surface, h_2 – rise distance from 0.5 m to 0.75m above stars surface.

$$\Delta m_4 \approx \frac{m_3 g_4 h_4}{c^2} \approx \frac{(9.05kg + 1.47kg) \times 4.404 \times 10^{16}\frac{m}{s^2} \times 0.25m}{(3 \times 10^8 \frac{m}{s})^2} \approx 1.29kg \quad \text{the mass increase}$$

for the final rise of 0.25m above star, $m_3 \approx 9.05$ kg+1.47kg$\approx$10.52kg - the mass at the 0.75m above surface, g_3-gravitational acceleration at 1 m above star surface, h_4 – rise distance from 0.75 m to 1 m above stars surface.

Overall mass increase is at least:

$$\Delta m \approx 2.30kg + 1.75kg + 1.47kg + 1.29kg \approx 6.81kg$$

This is better assessment for the mass increase then the one before, and as I mentioned you could use this method to find a much more precise result, you just need to slice elevation into more different increments.

Calculations are precise enough for our demonstration; we did see that the mass of the cat that lives on the surface of the neutron star-like object will increase more than two times for a distance of 1m.

m=5 kg   - the cat's mass at the surface of the neutron star-like object.

m'=m+$\Delta m \approx$ 5 kg+6.81kg$\approx$11.81kg –the cat's mass one meter above the named surface.

This is dramatic result that can be observed at extreme objects in our Universe, that is a region close to the $R \approx \frac{2G(m_1+m_2)}{c^2}$ and under conditions $\frac{2G(m_1+m_2)}{c^2} < R$ (again, in this region a gravitational force is attractive with large values as R goes near $\frac{2G(m_1+m_2)}{c^2}$), here m_1-cats mass, m_2-objects mass etc.

Note that amount of energy needed to perform this cat elevation would be more than enough to take Mount Everest and throw it one kilometer high; here at Earths conditions.

Max height for the pitch $H \approx \frac{\Delta m \times c^2}{M_{everest} \times g} \approx \frac{6.81 kg \times (3 \times 10^8 \frac{m}{s})^2}{5.75 \times 10^{13} kg \times 9.81 \frac{m}{s^2}} \approx 1161\ m$

6 C. Free fall in gravitational field and kinetic energy

Now would be a right moment for a reader to read again first 8-9 sentences in "Energy conservation in free fall".

You will increase a potential energy of the object if you raise it in the gravitational field from point 1 to point 2; in our story, with examples, so far, we have shown that the rest mass also increases in that process, at the same distance, in a way that:

(The rest mass difference between points 2 and 1)×(Light speed squared) =Increase in potential energy between points 1 and 2.

If you now drop the named object, in a free fall from point 2 to point1 a potential energy will decrease, but object's kinetic energy will increase in the same amount.

The decrease in a potential energy (from point 2 to point1) =The increase in kinetic energy (from point 2 to point 1)

To recapitulate, you will increase the potential energy and the rest mass of the object while you uplift it for the distance H (from point 1 to 2); and after that if you drop it (at point 2), an energy will remain constant in a free fall in a way that elevated potential energy acquired by raise will transform itself into the same amount of kinetic energy (at point 1), after it traverses H.   Or: a surplus of potential energy, that we created, at point 2, will completely go into surplus of kinetic energy at point 1, if we raise the body from point 1 to point 2, and let it fall from point 2 to point 1.

For weak fields: Example the Earth's surface:

$$\Delta PE \approx m_2 g H \approx \Delta m c^2 \approx (m_2 - m_1)c^2 = KE \approx \frac{m_1 v_1^2}{2} \quad (*)$$

ΔPE- The increase of the potential energy of the body, when we raise it from point 1 to 2.

m_2-the rest mass of the object at point 2, after we raise it for distance H

m_1-the rest mass of the object at point 1,

v_1-speed at point 1(speed at point 2 is zero)

KE-kinetic energy at point 1

For strong field, the object will have speeds comparable with speed of light, and if we assume as correct Einstein's equation for kinetic energy, at point 1 [14] speed of the falling object will be:

$$\Delta PE = \Delta m c^2 = (m_2 - m_1)c^2 = KE = \frac{m_1 c^2}{\sqrt{1-\frac{v_1^2}{c^2}}} - m_1 c^2 \quad (**)$$

ΔPE- The increase of the potential energy of the body, when we raise it from point 1 to 2.

m_2-the rest mass of the object at point 2, after we raise it for distance H

m_1-the rest mass of the object at point 1,

v_1-the speed at point 1(the speed at point 2 is zero)

KE-the kinetic energy at point 1

(Non-relativistic approximation; weak fields or small speed approximation - Maclaurin series for $v_1 \ll c$

$$m_2 c^2 = \frac{m_1 c^2}{\sqrt{1-\frac{v_1^2}{c^2}}} = m_1 c^2 \left(1 + \frac{1}{2}\left(\frac{v_1}{c}\right)^2 + \frac{3}{8}\left(\frac{v_1}{c}\right)^4 + \cdots\right) \approx m_1 c^2 + m_1 c^2 \frac{1}{2}\left(\frac{v_1}{c}\right)^2 \text{ therefore:}$$

$$m_2 c^2 - m_1 c^2 \approx \frac{m_1 v_1^2}{2}$$

On the other hand, here on Earth etc:

$$m_2 c^2 - m_1 c^2 \approx m_1 g h)$$

Example 10: Calculate the rest mass change in a free fall when speed reaches 0.866 of the speed of light.

Answer: We will use the equation (**).

$\Delta mc^2 = KE$

$$(m_2 - m_1)c^2 = \frac{m_1 c^2}{\sqrt{1 - \frac{v_1^2}{c^2}}} - m_1 c^2$$

$$m_2 c^2 - m_1 c^2 = \frac{m_1 c^2}{\sqrt{1 - \frac{v_1^2}{c^2}}} - m_1 c^2$$

$$m_2 c^2 = \frac{m_1 c^2}{\sqrt{1 - \frac{v_1^2}{c^2}}}$$

$$\frac{m_2 c^2}{m_1 c^2} = \frac{1}{\sqrt{1 - \frac{v_1^2}{c^2}}}$$

$$\frac{m_2}{m_1} = \frac{1}{\sqrt{1 - \frac{v_1^2}{c^2}}}$$

$\frac{m_2}{m_1} = \frac{1}{\sqrt{1 - \frac{(0.866c)^2}{c^2}}} \approx 2$ The rest mass becomes 2 times smaller when body reaches 0.866

of the speed of light (0.866C ≈ 2.6× $10^8 \frac{m}{s}$) .

Or, in percents of the original mass:

$\Delta m = m_2 - m_1 = m_2 - \frac{m_2}{2} = \left(1 - \frac{1}{2}\right)m_2 = \frac{1}{2}m_2 = \frac{50}{100}m_2$ This means that 50% of the original rest mass is lost in achieving the speed of 0.866C, in a free fall.

Example 11: Calculate the rest mass change in a free fall when speed reaches 0.9999 of the light speed, (99.99 % of the light speed).

Answer: Again, we will use the equation (**).

24

$\Delta mc^2 = KE$ ...

$\frac{m_2}{m_1} = \frac{1}{\sqrt{1-\frac{v_1^2}{c^2}}} = \frac{1}{\sqrt{1-\frac{(0.9999c)^2}{c^2}}} \approx 70.71$ The rest mass becomes approximately 70 times

smaller when the body reaches 0.9999 of the light speed ($0.9999C \approx 2.9997 \times 10^8 \frac{m}{s}$).

Or, in percents of the original mass:

$\Delta m = m_2 - m_1 = m_2 - \frac{m_2}{70.71} = \left(1 - \frac{1}{70.71}\right) m_2 \approx 0.986 m_2 = \frac{98.6}{100} m_2$ This means that
98.6% of the original rest mass is lost in achieving the speed of 0.9999C, in a free fall.

As we have seen in the past two examples, in a free fall you can achieve the speeds up to a speed of light, but you have to endure a very strong gravitational field and you will lose almost all of your rest mass.

Let us assume that you are that body that falls in the example 9, and that your initial rest mass is 70.71 kg. While you fall toward that very dense and massive celestial body, your speed will increase while, simultaneously, your rest mass decreases. At the end of the fall you lose 98.6% of original rest mass, your speed is 0.9999C relative to the starting point, and you have only 1 kg of the rest mass; in other words, if something stops you there, you will have only 1 kg; and you will transfer the rest of the energy (70.71kg-1kg)$\times c^2$ to the one that stopped you.

Example 12: (Weak field example) Apple falls from a coffee table to a floor, and it hits it with a speed of $3\frac{m}{s}$, calculate the apples' mass difference on the table and on the floor.

Answer:

$$v_{1=3}\frac{m}{s} \approx 3\frac{m}{s} \times \frac{C}{C} \approx 3\frac{m}{s} \times \frac{C}{3 \times 10^8 \frac{m}{s}} \approx 10^{-8}C$$

We can also use the equation (**) .

$\Delta mc^2 = KE$ ...

(Maclaurin Series [15];for x close to a zero holds: $\frac{1}{\sqrt{1-x^2}} \approx 1 + \frac{x^2}{2} + \frac{3x^4}{8} + \frac{5x^6}{16} + \cdots$)

$$\frac{m_2}{m_1} = \frac{1}{\sqrt{1 - \frac{v_1^2}{c^2}}} = \frac{1}{\sqrt{1 - \frac{(10^{-8}C)^2}{c^2}}} = \frac{1}{\sqrt{1 - (10^{-8})^2}} \approx 1 + \frac{(10^{-8})^2}{2} \approx 1 + 0.5 \times 10^{-16}$$

(Maclaurin Series [15]; for x close to a zero holds: $\frac{1}{1+x} \approx 1 - x + x^2 - x^3 + \cdots$)

$$\Delta m = m_2 - m_1 = m_2 - \frac{m_2}{1 + 0.5 \times 10^{-16}} = \left(1 - \frac{1}{1 + 0.5 \times 10^{-16}}\right) m_2 \approx$$

$$(1 - (1 - 0.5 \times 10^{-16}))m_2 = 0.5 \times 10^{-16} m_2 = \frac{5 \times 10^{-15}}{100} m_2 = \frac{0.000000000000005}{100} m_2$$

This means that the apple loses 0.000000000000005 % of its original mass when it falls from a coffee table.

6 d. A loss of the electric charge in extremely strong gravitational fields

In extremely strong gravitational fields a particle can lose significant part of its mass, under that circumstances we cannot expect that a particle can retain its electrical charge intact. Around charged particles there is an electrical field, that field have energy density, most of this energy is packed relatively close to the particle and it has to contribute to the mass of the particle.

6 e. A photon sphere radius

A photon also as another matter creates and feels gravitational force, so we can asses a radius R_F at which a photon rotates in circular orbit around a more massive object M. The energy of the photon is much smaller than the Massive object[14] so the expression for gravitational acceleration is:

$$a = G\frac{M}{R^2}\frac{1}{1 - \frac{2GM}{RC^2}}$$ (Note that M represents measurement of all energies encircled

by radius R).

For a photon radius R=R_F, the gravitational acceleration $a = a_F$ is a centripetal acceleration needed to hold a photon in a circular orbit.

$$a_F = \frac{c^2}{R_F}$$ A centripetal acceleration

---

[14] and radius is not close to 2GM/c^2

$$\frac{C^2}{R_F} = G\frac{M}{R_F^2}\frac{1}{1 - \frac{2GM}{R_F C^2}}$$

$$\frac{C^2}{R_F}(1 - \frac{2GM}{R_F C^2}) = G\frac{M}{R_F^2}$$

$$\frac{C^2}{R_F} - \frac{2GM}{R_F^2} = G\frac{M}{R_F^2}$$

$$\frac{C^2}{R_F} = \frac{3GM}{R_F^2}$$

$$R_F = \frac{3GM}{C^2} = \frac{3}{2}R_C$$

R_F- the radius of the photon path, which moves around the mass M with the speed c.

Among other things, this result means that a gravitational force is strong enough to hold a photon, under some circumstances.

Note similarity with "photon sphere" in the Schwarzschild calculations [16], with one significant difference, my expression for the gravitational force and acceleration have an area with its repulsive properties, and each and every of my equations, examples or calculations do not support violation of energy conservation in any shape, size or form.

An interesting observation: if we use the Newton's gravity equation, we will get three times smaller value for photon sphere radius.

7. A photonic structure of elementary particles

There are attempts [17] to explain the electron as some sort of intricate, self-entangled photon. Along with that, it is possible to create a particle-antiparticle pair from just a photon's "Electron-Positron Pair Production [18]" and it is also possible to annihilate the two particles and as a result to have just a photon's "Positron Annihilation [19]"

It is not too much stretch of imagination to observe the particles as some kind of the trapped photons. ("Electron-Positron Pair Production" as a photon - trapped, self-entangled. "Positron Annihilation"-a photon is untangled).

8. The Newton's and my own gravitational force comparison

Masses, distances and constants in both the Newton's and my own gravitational equation are always positive. The Newton's force is always attractive, and here is noted as positive[15]. My own equation is here noted as a positive one and attractive in some regions, while in the other regions it is negative and repulsive (with the same notation).

Let us consider a comparison between the Newtonian gravitational force and my own gravitational force, in series of examples we will observe two masses m and M in a free fall one to another and we will only consider forces' relation in certain regions, we will not go into the motion details at this point.

The Newton's expression for the gravitational force between the masses m and M at the distance R is: $F_N = G \frac{mM}{R^2}$

My own expression for the gravitational force between the masses m and M at the distance R is:

$$F = G \frac{mM}{R^2} \frac{1}{1 - \frac{2G(M+m)}{RC^2}}$$ In a free fall (isolated motions m and M one to other)

$$\frac{2G(M+m)}{c^2} = constant = R_c$$

At the distance R_c the force F is not defined, recall that F is defined for $0 < R < R_c$ and here F in my own equation is noted negative and therefore repulsive, and for

$R_c < R < +\infty$, F in my own equation is positive and therefore attractive.

Although the rest mass changes in motion; m and M do not, it is the invariant mass; in a free fall it is the measurement of energies before a motion, which is constant in isolated systems.

Again, for the isolated relation between the two bodies with the initial rest masses m and M; $\frac{2G(M+m)}{c^2} = constant = R_c$ always holds, even though their rest masses and relative speeds can change. So, for this case my gravitational force equation can be written as:

---

[15] See example 5.a.

$$F = G\frac{mM}{R^2}\frac{1}{1 - \frac{R_c}{R}}$$

Let us name R_c as a critical radius, it is a constant for the isolated systems, it has dimensions in meters, and again, the gravitational force F has the attractive properties for $R > R_C$, the gravitational force is not defined for $R = R_C$ and the gravitational force F has the repulsive properties for $R < R_C$.

a) At what distances of R in relation to R_C, the Newtonian gravitational force is almost equal to ours $(\frac{F_N}{F} \approx 1)$?

$$\frac{F_N}{F} = \frac{G\frac{mM}{R^2}}{G\frac{mM}{R^2}\frac{1}{1 - \frac{R_c}{R}}} = 1 - \frac{R_c}{R}$$

$\frac{F_N}{F} = 1 - \frac{R_c}{R} \approx 1$, only at$\frac{R_c}{R} \to 0$, in other words only for $R \gg R_C$ (R much larger than R_C) the Newtonian and our own gravitational force are almost equal.

b) At what distances of R in relation to R_C, the Newtonian gravitational force is almost equal to ours $(\frac{F_N}{F} = 0.9999999)$?

$$\frac{F_N}{F} = 1 - \frac{R_c}{R}$$

$$0.9999999 = 1 - \frac{R_c}{R}$$

$$\frac{R_c}{R} = 1 - 0.9999999$$

$$R = \frac{R_C}{1 - 0.9999999} = \frac{R_C}{0.0000001} = 10000000R_C$$

When the body with the mass m falls freely toward other body with the mass M, and they reach the distance of $10000000R_C$ the Newton's gravitational force is 99.99999 % the same as ours.

c) At what distances of R in relation to R_C, the Newtonian gravitational force is$\frac{F_N}{F} = 0.9999$?

29

Similar calculation yields: When they reach distance of R=10000R$_C$ the Newton's force is 99.99% of our gravitational force.

d) At what distances of R in relation to R$_C$, the Newtonian gravitational force is $\frac{F_N}{F}$ = 0.9?

Similar calculation yields: When they reach distance of R=10R$_C$ the Newton's force is 90% of our gravitational force.

e) At what distances of R in relation to R$_C$, the Newtonian gravitational force is $\frac{F_N}{F}$ = 0.5?

Again similarly, at R=2R$_C$, the Newton's gravitational force is two times weaker than ours.

f) At what distances of R in relation to R$_C$, the Newtonian gravitational force is $\frac{F_N}{F}$ = 0.1 (here $\frac{F}{F_N} = \frac{1}{0.1} = 10$), in other words, for what R=const*R$_C$, is our own force 10 times stronger than in Newton's equation?

$$\frac{F_N}{F} = 1 - \frac{R_c}{R}$$

$$0.1 = 1 - \frac{R_c}{R}$$

$$R = \frac{R_C}{1 - 0.1} = \frac{R_C}{0.9} = 1.11R_C = R_C + 0.11R_C$$

In your free fall, when you reach the distance that is 11% greater than R$_C$, you will experience the gravitational attractive force that is 10 times greater than the one predicted by the Newton's gravitational equation.

g) At what distances of R in relation to R$_C$, the Newtonian gravitational force is $\frac{F_N}{F}$ = 0.000001 (here $\frac{F}{F_N} = \frac{1}{0.1} = 1000000$), in other words, for what R=const*R$_C$, is our own force 1000000 times stronger than in the Newton's?

Similarly:

$$R = \frac{R_C}{1 - 0.000001} = \frac{R_C}{0.999999} = 1.000001R_C = R_C + 0.000001R_C$$

In your free fall, when you reach distance that is only 0.0001% greater than R_C, you will experience gravitational attractive force that is a million times greater than the one predicted by the Newton's gravitational equation.

h) At what distances of R in relation to R_C, the Newtonian gravitational force is $\frac{F_N}{F} = 10^{-40}$ (here $\frac{F}{F_N} = \frac{1}{10^{-40}} = 10^{40}$), in other words, for what R=const*R_C, is our own force 10^{40} times stronger than in Newton's?

(Maclaurin Series [15]; for x close to zero holds: $\frac{1}{1-x} \approx 1 + x + x^2 + x^3 + \cdots$)

$$R = \frac{R_C}{1 - 10^{-40}} = R_C \left(\frac{1}{1 - 10^{-40}} \right) \approx R_C(1 + 10^{-40} + (10^{-40})^2 + \cdots)$$
$$\approx R_C(1 + 10^{-40}) = R_C + R_C 10^{-40}$$

At $R \approx R_C + R_C 10^{-40}$, R is very close to R_C, and our own F is 10^{40} times greater than the one predicted by the Newton's gravitational equation.

In these regions the gravitational force expressed with our own relation is comparable with the electric forces at the same distances.

i) At what distances of R in relation to R_C, the Newtonian gravitational force is $\frac{F_N}{F} = 10^{-100}$ (here $\frac{F}{F_N} = \frac{1}{10^{-100}} = 10^{100}$), in other words, for what R=const*R_C, is our own force 10^{100} times stronger than in the Newton's equations?

Similar calculation to previous yields:

At $R \approx R_C + R_C 10^{-100}$, R is very, very close to R_C, and our own F is 10^{100} times greater than the one predicted by the Newton's gravitational equation.

In these[16] regions of R, the gravitational attractive force is stronger than any other that we could think of. In other words, in these regions the gravitational force is the absolute tyrant.

j) Note that the Newton's gravitational equation gives only an attractive, here noted as positive result, while my own equation will give positive results-attractive configuration for R>R_C and for R<R_C it will have a negative values-repulsive configuration.

At what distances of R in relation to R_C, the Newtonian gravitational force is related to ours as: $\frac{F_N}{F} = \frac{10^{-100}}{-1}$ (here $\frac{F}{F_N} = \frac{-1}{10^{-100}} = -10^{100}$), in other words, for what R=const*R_C, is our own force 10^{100} times stronger than in Newton's, and is our own force negative-repulsive?

$$\frac{F_N}{F} = 1 - \frac{R_c}{R}$$

$$\frac{10^{-100}}{-1} = 1 - \frac{R_c}{R}$$

$$\frac{R_c}{R} = 1 + 10^{-100}$$

(Maclaurin Series [15]; for x close to zero holds: $\frac{1}{1+x} \approx 1 - x + x^2 - x^3 + \cdots$)

$R = \frac{R_c}{1+10^{-100}} \approx R_C(1 - 10^{-100} + (10^{-100})^2 - \cdots) \approx R_C - R_C 10^{-100}$

$R \approx R_C - R_C 10^{-100}$, R is very, very close to R_C and just a little smaller, and our F is 10^{100} times greater than the one predicted by the Newton's gravitational equation, and it is repulsive.

Again, in this region where R is very, very close to R_C, and R<R_C, the gravitational force is repulsive and nothing is stronger than that.

k) At what distances of R in relation to R_C, the Newtonian gravitational force is $\frac{F_N}{F} = -0.5$?

---

[16] and even further than this.

$$0.5 = 1 - \frac{R_c}{R}$$

$$R = \frac{R_C}{1 + 0.5} = \frac{R_C}{1.5} = \frac{2}{3}R_C = R_C - \frac{1}{3}R_C \approx R_C - 0.33R_C$$

At this region R<R_C the gravitational force is repulsive, and smaller by a third of what would be the attractive Newton's gravitational force.

I will leave up to a reader to envision what can happen to our objects with the masses m and M, if they can pass through all of our points calculated in the previous examples. Note only that the energy conservation is a must, and that your rest mass, in a free fall, can be reduced up to almost a zero; while your speed can increase up to almost speed of light.

9. The electric force versus the gravitational force

In the current state of affairs the electric force is, always, much stronger than the gravitational force. Contrary to that my own equation shows that there are regions were the gravitational force is equal to electric forces and near to that region is the region where the gravitational forces are much stronger then the electric forces. In standard physics electric forces can be repulsive (for same electric charges) or attractive (for opposite electric charges) while the gravity is always attractive. In my point of view the electric forces are as they are, but the gravity can also be repulsive, see: 8j, 8k, the example 5a, etc.

To illustrate this relation, there is one recurrent good example widely used[17], it usually goes something like:

Example 13: "Suppose you put two protons a certain distance apart. They would be attracted by a gravitational force and repelled by an electric force. **Is there a distance at which these two forces would balance?**"

Standard (not my own) answer: No, because

---

[17] Some version of it: "Electric Forces Versus Gravitational Forces"-
http://www.batesville.k12.in.us/physics/PhyNet/e%26m/electrostatics/michaels_question.htm
"6. Determine the ratio of the electric force to the gravitational force between a proton and an electron."-
http://sprott.physics.wisc.edu/phys104/sol1s02p.pdf

$$\frac{\text{Electric force}}{\text{Newton's gravitational force}} = \frac{F_E}{F_G} = \frac{K\frac{q_p q_p}{R^2}}{G\frac{m_p m_p}{R^2}} = \frac{K q_p q_p}{G m_p m_p} = \frac{K(q_p)^2}{G(m_p)^2}$$

$$= \frac{9 \times 10^9 \frac{Nm^2}{C^2}(1.6 \times 10^{-19}C)^2}{6.67 \times 10^{-11}\frac{Nm^2}{kg^2}(1.67 \times 10^{-27}kg)^2} \approx 1.24 \times 10^{36}$$

$\frac{F_E}{F_G} \approx 10^{36}$ According to standard version, the relation of the electric force and the gravitational force for two protons is constant and the electric force is always 10^{36} times greater than the gravitational one.

My own answer: **Yes**, in theory (and it is possible in reality, but not for two protons):

$$1 = \frac{\text{Electric force}}{\text{My gravitational force}}$$

$$1 = \frac{F_E}{F} = \frac{K\frac{q_p q_p}{R^2}}{G\frac{m_p m_p}{R^2}\frac{1}{1-\frac{2G(m_p+m_p)}{RC^2}}} = \frac{K q_p q_p}{G m_p m_p}\left(1 - \frac{2G(m_p+m_p)}{RC^2}\right) = \frac{K(q_p)^2}{G(m_p)^2}\left(1 - \frac{4G m_p}{RC^2}\right)$$

$$= \frac{9 \times 10^9 \frac{Nm^2}{C^2}(1.6 \times 10^{-19}C)^2}{6.67 \times 10^{-11}\frac{Nm^2}{kg^2}(1.67 \times 10^{-27}kg)^2}\left(1 - \frac{4 \times 6.67 \times 10^{-11}\frac{Nm^2}{kg^2} \times 1.67 \times 10^{-27}kg}{R\left(3 \times 10^8 \frac{m}{s}\right)^2}\right) \approx$$

$$1.24 \times 10^{36}(1 - \frac{4.95 \times 10^{-54}m}{R})$$

$1 \approx 1.24 \times 10^{36}(1 - \frac{4.95 \times 10^{-54}m}{R})$ Solution for R (note that here for R=R$_C$=4.95 × 10^{-54}m function F is not defined)

$$\frac{4.95 \times 10^{-54}m}{R} \approx 1 - \frac{1}{1.24 \times 10^{36}}$$

$$R \approx 4.95 \times 10^{-54}m \times \frac{1}{1 - \frac{1}{1.24 \times 10^{36}}}$$

(Maclaurin Series [15]; for x close to a zero holds: $\frac{1}{1-x} \approx 1 + x + x^2 + x^3 + \cdots$)

$$R \approx 4.95 \times 10^{-54}m(1 + \frac{1}{1.24 \times 10^{36}} + \left(\frac{1}{1.24 \times 10^{36}}\right)^2 + \left(\frac{1}{1.24 \times 10^{36}}\right)^3 \ldots)$$

$$R \approx 4.95 \times 10^{-54}m(1 + \frac{1}{1.24 \times 10^{36}})$$

$R \approx 4.95 \times 10^{-54}m + 0.8 \times 10^{-36} \times 4.95 \times 10^{-54}m$ -Distance, at which the gravitational force is equal to the electrical force, for two protons.

Or$R \approx R_C + 0.8 \times 10^{-36}R_C$, Result for the distance R is as expected (se 8.h. pg. 27,28), just above what we named the critical radius. Remember that, mathematically, you could place them even closer which would result in even greater values for the gravitational force (see 8.i. pg.28)

Be that as it may, it is unrealistic to assume that the centers of the two protons can come that close together simply because a size of a proton is somewhere around $10^{-15}m$[20] and we calculated the equilibrium distance to be around $10^{-54}m$, and that is well inside of a proton.
Not to be discouraged, I strongly believe and I will try to convince you as well, that the balance between the attractive gravitational force, on one side, and the electromagnetic repulsion, on the other, is something that defines one and only building block of this nature.

Example 14: Calculate equilibrium distance between the electric and the gravitational forces for two equal electric charges of

$$q_{\frac{d}{2}} = \frac{1}{2}(of\ d\ quark\ electric\ charge) = \frac{1}{2} \times \frac{1}{3} \times (-1.6 \times 10^{-19}C) \approx 2.67 \times 10^{-20}C$$

With two equal masses off

$$m_{\frac{d}{2}} = \frac{1}{2}(mass\ of\ d\ quark\ in\ proton^{18}) = \frac{1}{2} \times 1.78 \times 10^{-30}kg \approx 8.9 \times 10^{-31}kg$$

Answer: From previous examples we know that the only place where the gravitational force is attractive and strong enough to balance the electric repulsive force is just a little more than the critical radius. Hence, the equilibrium distance is:

---

[18] If we accept as correct [21]

$$R \approx \frac{2G\left(m_d + m_d\right)}{C^2} = \frac{4Gm_d}{C^2} \approx \frac{4 \times 6.67 \times 10^{-11}\frac{Nm^2}{kg^2} \times 8.9 \times 10^{-31}kg}{(3 \times 10^8 \frac{m}{s})^2}$$

$$\approx 2.64 \times 10^{-57} m$$

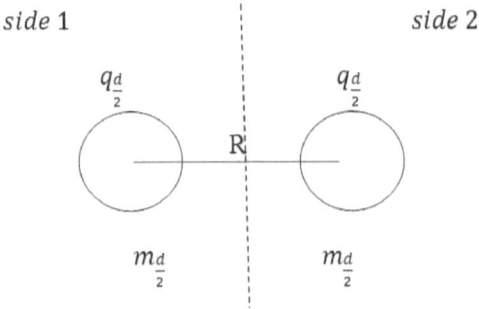

These two identical electric charges $q_{\frac{d}{2}}$, with two identical masses $m_{\frac{d}{2}}$, are in balance at the distance R. They are attracted to each other with the gravitational force and separated by an electric force.

10.   The quark-ed structure of electrons (unit charge $= e = 1.6*10^{-19}$ C)

According to "standard model of elementary particles" a proton has one unit of a charge: "But a proton with one unit of charge is composed of three quarks - two up quarks and one down quark. To make this work, it turns out that the up quark has a charge of +2/3 of the unit charge and the down quark has -1/3 units of charge. So the sum of the charges is +1 unit, whereas the electron has -1 unit of charge."[22]

I see no other reason for an electron to have exactly three times the electric charge of d quark, but to be a composite particle made of three d quarks. And in that line of thought; because the up quark has exactly two times the electric charged of anti-down quark, and therefore:

It is my personal opinion that:

    a.   An electron consists of three down quarks.

    b.   The up quark consists of two anti-down quarks

etc.

    c.   The difference in mass does not mean a new quark, but rather the same quark in different interaction. (In other words; down, strange and bottom quarks are the same particles with a different mass caused by different interactions. Same holds for: up, charm and top quark etc.)

Keep in mind that it is possible to have places in this nature where the anti-proton mass is the same as the mass of electron here on the surface of the Earth.

Important note: Electron-quarked structure hypothesis is a long shot, but it is my strong persuasion that we need to search in that general direction; if not three d quarks in the electron, maybe there are 13 or 133 or who knows how many. We need to open our minds to all possibilities, in quark-ed unusual world, a protonic assembly of quarks can easily have 1836 times bigger mass than assembly 133 d (and 130 anti-d) quarks.

11. Down and anti-down quark structure.

What did we see so far? We invoke the logic to conclude that: if you, as a separate object, have the rest mass; you are some sort of combination of d and anti-d quarks, there is no object that is made of something else (10.). We calculated the distance at which the electric attractive force is balanced with very strong gravitational force; it was the region near R_C (example 14.). We concluded that particles can be some sort of photonic assembly (7.), but photons have balanced electric charges[19], while our d and anti-d quarks have surplus in negative and positive electric charges respectively, but then we remembered, that we can't expect particle to hold on to all of its electric charge in a very strong gravitational field i.e. near R_C (6.d.).

All in all, d and anti-d quark should be: a sort of a mirror image to each other, photonic in its nature, extremely dense; R close to R_C, they have misbalance in its own

---

[19]whatever electric charge might be

37

respective electric charges, and they are expected to have dipole magnetic moments so they should have sort of circular motion of misbalanced electric charges, but not real circular motion because that would radiate electromagnetic energy away. Having all of this in mind:

Simplified representation of a photon, d quark and anti-d quark:

Photon (more or less standard representation)

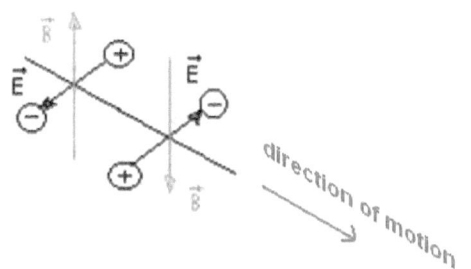

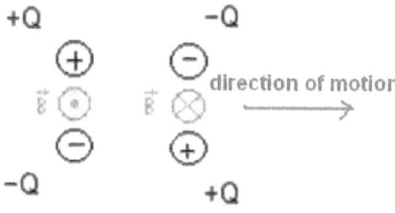

d and anti-d quark (my personal construction)

quark, view from above

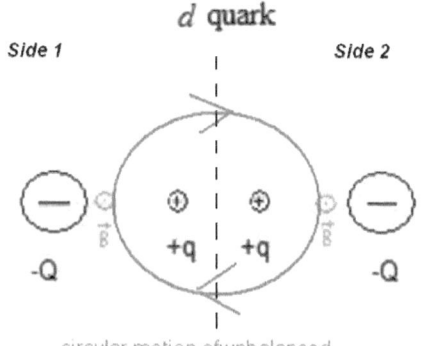

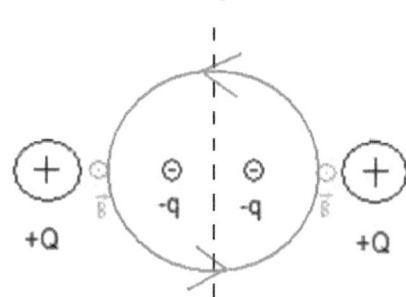

circular motion of unbalanced electrical charges

Electric charge of the d quarq

$$q_d = 2 \times (-Q) + 2 \times (q) = -\frac{1}{3} e = -\frac{1}{3} 1.6 \times 10^{-19} C$$

circular motion of unbalanced electrical charges

Electric charge of the anti-d quarq

$$\bar{q}_d = 2 \times (Q) + 2 \times (-q) = \frac{1}{3} e = \frac{1}{3} 1.6 \times 10^{-19} C$$

$$|+Q| = |-Q|$$
$$|+q| = |-q|$$
$$|\pm Q| > |\pm q|$$

$$e = 1.6 \times 10^{-19} C$$

In the spirit of all of the above, we can assume a d quark structure as a twisted, mutilated, self-entangled "photon" held together and as it is with its own extremely strong gravitational field (R> R$_C$) and held apart by the electromagnetic repulsion; in addition, bear in mind that in its strongest domain (R< R$_C$ and very, very close to R$_C$) a gravitational repulsive force has no match, it is an unchallenged ruler.

In a picture of a d quark (picture above) we see four electric charges, and they are: two external negative electric charges Q and two internal positive electric charges q, exact quantity of Q and q in Coulombs is unknown to me at this time, but their overall sum is exactly one third of a negative elementary charge $e \approx 1.6 \times 10^{-16} C$ [23], see picture above for reference and calculation. The internal electric charges are subjected to a greater (than external) rest mass loss and as a result its quantity is reduced more (see 6.d.), in reality it is more likely that its electric field is the one that gets reduced, but we observe it as an electric charge reduction. If we observe the named quark from a big enough distance, that starts somewhere around 10^{-14} m[20] (for d quark within a proton), we can't detect these internal positive electric charges within a d quark, at these and larger distances we observe d quark as one particle with one electric charge of –e/3, but other particles that are close enough, they certainly can. These almost[21] planar imbalanced electrical charges within one d quark[22] will enable us to consider nuclear and color forces as an electromagnetic interaction.

The first scenario:

In the example 14 you can find an assessment for diameter R (very, very near 2×critical radius R$_C$) within a d quark at which the gravitational attractive force is balanced with an electromagnetic repulsion, as we calculated it is in region:

$$R \approx 2.64 \times 10^{-57} m$$

To asses magnitude of R in the d quark in the proton, I modeled entire "side 1" in the pictures of the d quark as one separate particle with mass $m_{\frac{d}{2}}$ and electric charge of $q_{\frac{d}{2}}$ and the entire "side 2" of the quark as another particle with the same charge and the same mass (compare pictures, note dotted line that separates "side 1" and "side 2"; "side 1" and "side 2" pictures are, of course, just for calculation and explanation).

---

[20] some distance larger than the nucleus diameter http://hyperphysics.phy-astr.gsu.edu/hbase/Nuclear/nucuni.html
[21] It is most likely that the positive electric charges in d quark (see picture) are somewhat above (or below) flat plane, i.e. d quark most likely have electric dipole properties (of different electric charges).
[22] Remember that the upper quark is an assembly of two anti-down quarks.

About d quark rest mass

Someone (I wasn't able to find who or when) posted a part of a story about a photon and the rest mass, in comments for a post in linked-in group "Theoretical physics"[23] it goes something like this: "Empty mirrored box has a smaller rest mass than the same box full of photons that bounce inside from mirror to mirror, even though a photon itself does not have the rest mass". It is similar situation with the d quark (and anti-d quark), as a part of another particles (proton, neutron, electron etc) it contributes to the rest mass (in a way as photons in a mirrored box), whether or not it has the rest mass on its own.

From time to time conditions meet for a photon's "positive electric charge" to become too close to its own "negative electric charge" (a photon bends to one side or the other) in that position the electric force between the two charges on one flank is bigger enough than the one on the other flank, so a photon starts to bend to one side even more, the attractive electric force becomes bigger and bigger and a photon bends farther, now in addition to the electric attractive force; the gravitational attractive force becomes also strong ($R > R_C$ and close to R_C), and now, both the electric and gravitational force take part in further photon bending...up to the point where energy density becomes very large and one part of a photon enters the region of the repulsive gravitational force ($R < R_C$ and very, very close to R_C) there this force rules; it slaps photon so hard that it gets twisted around its axes (one way or another[24]) and settles down twisted as in the picture above. Now, again from time to time, it tries to become as a normal photon in the gravitational orbit, but, you guess it right, it gets slapped back to its self-twisted position, again.

In reality, d and anti-d quarks are always created as a part of other particles, and every point of our space is capable of their production under known circumstances. The story above is just a nice visualization (but not without real importance), unfortunately things are never this simple.

(An unrelated thought- there is a nonzero possibility that charges within a d quark similar to d quark itself, just with smaller parts, and so on. Or, d and anti-d quarks are in fact deformed photon-like particles...but 4 imbalanced electric charges in a photon can

[23]Theoretical Physics, (Eddie A. Maalouf group owner) https://www.linkedin.com/groups/3091009
[24] d quark or anti-d quark.

be smaller d and anti-d quarks, and so on indefinitely to infinitesimally small photon or a quark[25]. It is healthier not to think in this direction for a time being.)

In the picture of a photon on page 34 we see one wavelength of a named particle in "straight motion"; in the picture of a d quark (and anti-d quark) we see one wavelength of photon-like particle in "circular motion". An important note: "circular motion" of the d quark is to be understood similarly to "straight" motion of the photon. A photon is a particle and it can move straight, but its movement is not completely as a straight flight of a pebble. Quark (down and anti-down) is a particle and it rotates, but its rotation is not completely as the rotation of the pebble, again circular motion of the d quark is to be understood as straight motion of the photon.

Example 15: Let us calculate some characteristic quantities for a d quark within a proton.

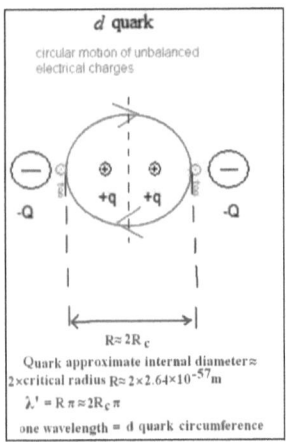

d quark

circular motion of unbalanced electrical charges

-Q   +q   +q   -Q

R≈ 2R꜀

Quark approximate internal diameter≈
2×critical radius R≈ 2×2.64×10⁻⁵⁷m

$\lambda' = R\pi \approx 2R_c\pi$

one wavelength = d quark circumference

An approximate internal diameter: $R \approx 2 \times 2.64 \times 10^{-57}m$ (see page 32.) this is distance at which the electric repulsive force is balanced with the gravitational attractive force, a real size of the particle is larger (see how much larger in scenario 3).

The rest mass of the d quark within a proton:
$$m_d \approx 1.78 \times 10^{-30}kg$$

The Plank's constant: let us calculate the value of the named constant h' at distance R, having in mind that the d quark is a photon-like particle, so its speed is $C = 3 \times 10^8 \frac{m}{s}$, its wavelength λ' and frequency f' are connected as C=λ 'f' and that it has energy of $E = m_d C^2$ as well as $E = h'f'$ therefore

$m_d C^2 = h'f'$ or $m_d C^2 = h'\frac{C}{\lambda'}$ and remember λ'is one circumference $\lambda' = R\pi \approx$ $2 \times 2.64 \times 10^{-57}m \times 3.14 \approx 1.66 \times 10^{-56}m$ hence:

$$h' = m_d C\lambda' \approx 1.78 \times 10^{-30}kg \times 3 \times 10^8\frac{m}{s} \times 1.66 \times 10^{-56}m \approx 8.86 \times 10^{-78}Js$$

Frequency: $f' = \frac{C}{\lambda'} \approx \frac{3\times10^8\frac{m}{s}}{1.66\times10^{-56}m} \approx 1.81 \times 10^{64}Hz$

---

[25] Similar to: What came first the chicken or the egg?

So far we assessed size, frequency, wavelength and the value of the Plank's constant in region of R, again, the actual size of the d quark in the proton has to be bigger. See calculation for R, and compare approximations with the help of pictures.

If our d quark could untangle itself (in reality quarks are crated and destroyed in bulks, it is more energy efficient that way, see scenario 3); become a free photon, and come to our laboratory, here on surface of the Earth, what kind of frequency and wave length would we measure?

In escaping his own very strong gravitational bonds, the energy will conserve (as would in any other case) E'=E.

Frequency: Here on surface of the Earth Plank's constant has value of $h = 6.626 \times 10^{-34} JS$ E'=E, $h'f' = hf$

$$f = \frac{h'}{h} f' \approx \frac{8.86 \times 10^{-78} Js}{6.626 \times 10^{-34} JS} 1.81 \times 10^{64} Hz \approx 2.42 \times 10^{20} Hz$$

Wavelength, here at the Earth's surface: $\lambda = \frac{C}{f} = \frac{3 \times 10^8 \frac{m}{s}}{2.42 \times 10^{20} Hz} \approx 1.24 \times 10^{-12} m$

Comment on results: Calculation showed us what we can expect in these extreme conditions, we acquire some feeling about magnitudes, but in reality, both positive and negative charges are more likely to be in regions beyond R_C, with positive charges closer to that region, in order to lose larger part of its rest mass, and its electric charge along with it. Negative charges have to be further from center, to avoid identical faith as positive internal electric charges (this holds for the d quark, for anti-down d quark is other way around), both charges can be reduced but the internal one has to be reduced more.

The second scenario:

More accurate d quark structure (d quark in a proton)

d quark is a photon-like particle that goes around itself, it is held by its own gravitational field, one wavelength, twisted, with reduced electric charges and positive-internal charges with greater charge reduction. d quark is a photon-like particle, so the radius of his rotation is in fact a photon sphere radius R_F and it rotates around half of its rest mass$\frac{m_d}{2}$. (In reality an electric field oscillates normally to the direction of motion (the direction of motion -red line in the picture below)) but synchronized in a way that a positive charge is always inside, it is progressive EM wave, so we perceive it as a charge that moves in circular orbit. In other words, real electric charges (if any) moves perpendicular to a circle, not in direction of a circle, but we observe that as a circular motion of Q and q). Note that timelines will differ, and that it is created this way, further on this at the end.

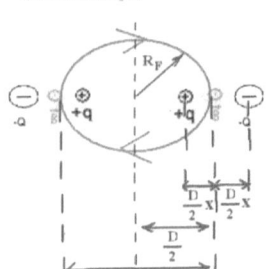

d quark

circular motion of unbalanced electrical charges

R_F

$-q$ $+q$ $+q$ $-Q$

$\frac{D}{2}$ x | $\frac{D}{2}$ x

$\frac{D}{2}$

$D=2R_F\pi$

$\lambda' = D\pi \approx 2R_F\pi$

one wavelength = d quark circumference

Photon sphere radius satisfies the following condition:

$$R_F = \frac{3}{2}R_C = \frac{3}{2}\frac{2GM}{C^2} = \frac{3GM}{C^2}$$

Here: $M = \frac{m_d}{2}$

$$R_F = \frac{3G\frac{m_d}{2}}{C^2} = \frac{3 \times 6.67 \times 10^{-11}\frac{Nm^2}{kg^2} \times 1.78 \times 10^{-30}kg}{2 \times (3 \times 10^8 \frac{m}{s})^2}$$

$R_F = 1.98 \times 10^{-57}m$ This is the radius of the red line around which electric and magnetic field oscillate (as a photon around direction of its motion); again the overall electric charge is negative e/3, and we can't detect internal charges if we stand far enough. The diameter of the particle: $D = 2R_F = 2 \times 1.98 \times 10^{-57}m = 3.96 \times 10^{-57}m$, wavelength of this photon-like particle (one circumference):

$$\lambda' = D\pi = 3.96 \times 10^{-57}m \times 3.14 \approx 1.24 \times 10^{-56}m$$

Frequency at those conditions: $f' = \frac{C}{\lambda'} = \frac{3\times10^8\frac{m}{s}}{1.24\times10^{-56}m} \approx 2.42 \times 10^{64}Hz$

Value of the Plank's constant at those conditions:

$$h' = \frac{m_d C^2}{f'} = \frac{1.78 \times 10^{-30}kg(3 \times 10^8 \frac{m}{s})^2}{2.42 \times 10^{64}Hz} \approx 6.675 \times 10^{-78}Js$$

Time (gravitational dilatation)

The value of the Plank's constant in our laboratory: $h = 6.626 \times 10^{-34} Js$

If we observe some photon that originated from the environment where the value of Plank's constant is $h' = 6.675 \times 10^{-78} Js$, in its voyage up to us, its frequency will decrease, but its energy will remain constant I.E.

E'=E

$$h'f' = hf \quad (f = \frac{h'}{h} f' \approx 2.42 \times 10^{20} Hz)$$

$$h'\frac{1}{T'} = h\frac{1}{T}$$

$$T' = \frac{h'}{h} T = \frac{6.675 \times 10^{-78} Js}{6.626 \times 10^{-34} Js} T \approx 10^{-44} T$$

We always perceive ourselves to be the same (the d quark probably thinks that it is just another completely normal photon), but if we find ourselves to be in the environment where the Plank's constant has the value of $6.675 \times 10^{-78} Js$, one of our second will correspond to 10^{44} seconds, here at the surface of the Earth.

To put things into perspective; some people say that our Universe is 4.32×10^{17} seconds old.

Length (gravitational contraction)

Wavelength of our photon $\lambda' \approx 1.24 \times 10^{-56} m$ in conditions where Plank's constant has the value of $h' = 6.675 \times 10^{-78} Js$, will have significantly larger value in our own laboratory.

E'=E

$$h'f' = hf$$

$$h'\frac{c}{\lambda'} = h\frac{c}{\lambda}$$

$$\lambda' = \frac{h'}{h} \lambda = \frac{6.675 \times 10^{-78} Js}{6.626 \times 10^{-34} Js} \lambda \approx 10^{-44} \lambda$$

$$\lambda = \lambda' \times 10^{44} = 1.24 \times 10^{-56}m \times 10^{44} \approx 1.24 \times 10^{-12}m$$

A photon will be significantly smaller in the environments where h' describes the processes, having in mind that the same holds for every other object in our reality, because you are either a photon or you are made of d and anti-d quarks (who are more or less photons), and there is nothing more. So, if a photon contracts, you will contract as well.

The third scenario (final and the most accurate one):

Let us asses the magnitude of the electric charges Q and q within one d quark, and their separation from the direction of motion $\frac{D}{2}$x, as well as the diameter D.

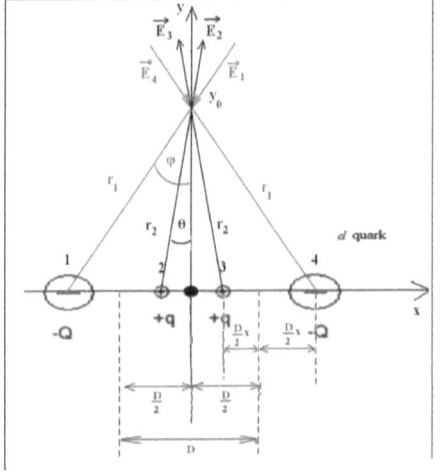

I will try to estimate the charges Q and q with the help of a nuclear force between a proton and a neutron within a deuteron. In this case, I will consider nuclear force to be an electromagnetic interaction of the d quark from a proton and the anti-d quark from the up quark within a neutron. (Calculations showed to me that the color and the nuclear force, as this kind of interaction[26], exist only between two predominantly different electrical charges with the same direction of rotation). My calculation will be somewhat imprecise because there are other charges in vicinity. This calculation is demonstrative, but not without a real value. Better assessments are possible and calculable with more data, interactions and better computers. It has been said that [32] Coulomb's force between two protons dominates at distances around 2.5 fm, and in our own perspective it means that this is a distance at which the two mentioned quarks from neighboring proton and neutron cannot see the internal positive (within protons d quark) and the internal negative charges (within a neutron up quarks anti-d quark) in each other. At those distances the electrical interaction is strictly attractive (between d and anti-d quark), as a contrast from shorter separations where electrical interaction has its

---

[26] Similar, but different are possible.

repulsive parts. At distances little less than 2.5 fm, the electrical field at y axis (see picture above, point $(0, y_0)$) within d quark is zero (if d quark has no positive internal charges electrical field will be negative at every point in surroundings of d quark, and positive at every point in its surroundings of anti-d quark. Therefore modeling a nuclear force as an electromagnetic interaction would be impossible). At points on y axis between the points $(0, y_0)$ and $(0, 0)$ and for between two symmetrical points on y axis, below x axis, $(0, 0)$ and $(0, -y_0)$ the electrical field is positive. At every distance larger than y_0, from the d quark electric field is negative. In other words you can't see the internal positive charges at a distance greater than y_0 (at y axis, other points off y axis are possible and calculable). For anti-d quark the story is similar, but reversed. In essence we have four unknown magnitudes Q, q, x and D (see the picture above), so we will use four equations. First is the obvious one $Q - q = \frac{1.6 \times 10^{-19}}{2 \times 3} \approx 2.667 \times 10^{-20} C$ -First equation-.

The second equation is from conditions for zero electric field at y axis, just before electric force domination i.e. right after d quarks positive internal charges become undetectable, let's say $y_0 \approx 2.5 fm \approx 2.5 \times 10^{-15} m$. Calculation similar to: "linear electric quadrupole" [33].

See picture above: $E_y = 0$ for:

$E = k \frac{Q}{r^2} \left[\frac{N}{C}\right]$ –The magnitude of the electric field at distance r from the electric charge Q.

$k \approx 9 \times 10^9 \frac{Nm^2}{C^2}$ Coulomb constant

$$E_{1x} = k\frac{Q_1}{r_1^2}\sin\varphi = k\frac{Q_1}{r_1^2}\frac{\frac{D}{2}(1+x)}{r_1} = k\frac{Q_1\frac{D}{2}(1+x)}{r_1^3}, E_{1y} = k\frac{Q_1}{r_1^2}\cos\varphi = k\frac{Q_1}{r_1^2}\frac{y_0}{r_1} = k\frac{Q_1 y_0}{r_1^3}$$

$$E_{4x} = k\frac{Q_4}{r_1^2}\sin\varphi = k\frac{Q_4}{r_1^2}\frac{\frac{D}{2}(1+x)}{r_1} = k\frac{Q_4\frac{D}{2}(1+x)}{r_1^3}, E_{4y} = k\frac{Q_4}{r_1^2}\cos\varphi = k\frac{Q_4}{r_1^2}\frac{y_0}{r_1} = k\frac{Q_4 y_0}{r_1^3}$$

$$E_{2x} = k\frac{q_2}{r_2^2}\sin\theta = k\frac{q_2}{r_2^2}\frac{\frac{D}{2}(1-x)}{r_2} = k\frac{q_2\frac{D}{2}(1-x)}{r_2^3}, E_{2y} = k\frac{q_2}{r_2^2}\cos\theta = k\frac{q_2}{r_2^2}\frac{y_0}{r_2} = k\frac{q_2 y_0}{r_2^3}$$

$$E_{3x} = k\frac{q_3}{r_2^2}\sin\theta = k\frac{q_3}{r_2^2}\frac{\frac{D}{2}(1-x)}{r_2} = k\frac{q_3\frac{D}{2}(1-x)}{r_2^3}, E_{3y} = k\frac{q_3}{r_2^2}\cos\theta = k\frac{q_3}{r_2^2}\frac{y_0}{r_2} = k\frac{q_2 y_0}{r_2^3}$$

Here $Q_1 = Q_4 = Q$, and $q_2 = q_3 = q$.

Due to symmetry, see picture: $E_{1x} = -E_{4x}$, and $E_{2x} = -E_{3x}$

$$E_{1x} + E_{2x} + E_{3x} + E_{4x} = 0$$

For E=0 in point $y_0 = 2.5 \times 10^{-15}m$, see picture, $E_{2y} + E_{3y} = E_{1y} + E_{4y}$, and $E_{2y} = E_{3y}, E_{1y} = E_{4y}$.

$$2k\frac{qy_0}{r_2^3} = 2k\frac{Qy_0}{r_1^3}$$

$$\frac{q}{r_2^3} = \frac{Q}{r_1^3}$$

$$\frac{Q}{q} = \frac{r_1^3}{r_2^3} = \frac{\left(\left(\frac{D}{2}(1+x)\right)^2 + y_0^2\right)^{\frac{3}{2}}}{\left(\left(\frac{D}{2}(1-x)\right)^2 + y_0^2\right)^{\frac{3}{2}}} = \left(\frac{D(1+x)^2 + (2y_0)^2}{D(1-x)^2 + (2y_0)^2}\right)^{\frac{3}{2}} = \left(\frac{D(1+x)^2 + (2\times 2.5\times 10^{-15}m)^2}{D(1-x)^2 + (2\times 2.5\times 10^{-15}m)^2}\right)^{\frac{3}{2}}$$

$$\frac{Q}{q} = \left(\frac{D(1+x)^2 + 2.5\times 10^{-29}m^2}{D(1-x)^2 + 2.5\times 10^{-29}m^2}\right)^{\frac{3}{2}} \text{ -The second equation-}$$

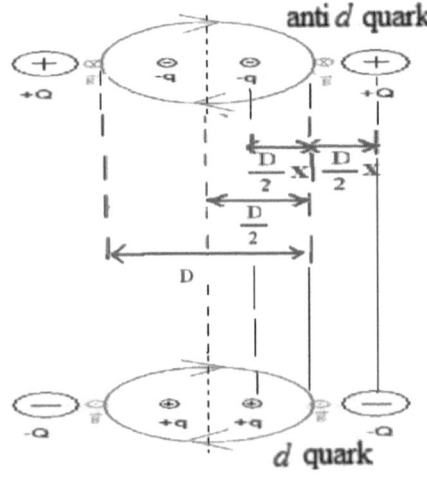

anti d quark

d quark

The third and the fourth equation is obtained from the electromagnetic force equation.

For d and anti-d quarks at separation y, overall approximate electrical attractive force [34] between two quarks; with origin at these eight electric charges is:

$$F_E = 4k\left(\frac{Q^2}{y^2} + \frac{q^2}{y^2} - \frac{2Qq}{y^2 + (Dx)^2}\right)$$

And for an overall magnetic interaction:

In general, the magnetic force between two parallel current loops at distances y>>a, and a≈ b (a, b-radiuses of the current loops) [35]:

$$F_M = \frac{3\,\mu\pi}{2}\frac{1}{y^4}a^2b^2 I_a I_b$$

I_a-electrical current of the current loop with radius a,

I_b -electrical current of the current loop with radius b

$\mu = 4\pi 10^{-7} \frac{Tm}{A} = 4\pi 10^{-7} \frac{N}{A^2}$ -magnetic constant

After some calculation and adaptation to our situation (repulsive magnetic force of predominantly opposite electric charges that rotate in the same direction) we get the magnetic force for our two quarks as:

$$F_M = \frac{3\,\mu\pi}{2}\frac{}{y^4}(R_Q^2 I_Q - R_q^2 I_q)^2$$

$$F_M = \frac{3\,\mu\pi}{2}\frac{}{y^4}(R_Q^2 \frac{2QC}{2R_Q\pi} - R_q^2 \frac{2qC}{2R_q\pi})^2$$

$$F_M = \frac{3\,\mu\pi}{2}\frac{C^2}{y^4}\frac{}{4\pi^2}(R_Q 2Q - R_q 2q)^2$$

$$F_M = \frac{3\,\mu}{2}\frac{C^2}{4\pi}\frac{D}{y^4}(\frac{D}{2}(1+x)2Q - \frac{D}{2}(1-x)2q)^2$$

$$F_M = \frac{3\,\mu}{2}\frac{C^2}{4\pi}\frac{}{y^4}(D(1+x)Q - D(1-x)q)^2 \quad \text{Or}$$

$$F_M = \frac{3\,\mu}{2}\frac{}{4\pi}C^2\frac{D^2}{y^4}(Q - q + (Q + q)x)^2$$

Only a stabile balance force in this kind of configuration is when: $F_{EM} = F_M - F_E$ (See the picture above). Therefore:

Equation for the nuclear force as an electromagnetic interaction is:

$$F_{EM} = \frac{3\,\mu}{2}\frac{}{4\pi}C^2\frac{D^2}{y^4}(Q - q + (Q + q)x)^2 - 4k(\frac{Q^2}{y^2} + \frac{q^2}{y^2} - \frac{2Qq}{y^2 + (Dx)^2})$$

Or

$$F_{EM} = \frac{3\,\mu}{2}\frac{}{4\pi}C^2\frac{D^2}{y^4}(Q - q + (Q + q)x)^2 - \frac{4k}{y^2}(Q^2 + q^2 - 2Qq\frac{y^2}{y^2 + (Dx)^2})$$

49

And because:

$$Q^2 + q^2 - 2Qq\frac{y^2}{y^2 + (Dx)^2} = Q^2 + q^2 - 2Qq + 2Qq - 2Qq\frac{y^2}{y^2 + (Dx)^2} =$$

$$(Q - q)^2 + 2Qq\left(1 - \frac{y^2}{y^2 + (Dx)^2}\right) = (Q - q)^2 + 2Qq\left(1 - \frac{y^2}{y^2 + (Dx)^2}\right) =$$

$$(Q - q)^2 + 2Qq\frac{(Dx)^2}{y^2 + (Dx)^2}$$

We can write same F_{EM} equation as:

$$F_{EM} = \frac{3}{2}\frac{\mu}{4\pi}C^2\frac{D^2}{y^4}(Q - q + (Q + q)x)^2 - \frac{4k}{y^2}((Q - q)^2 + 2Qq\frac{(Dx)^2}{y^2 + (Dx)^2})$$

We know that nuclear force [36] has its zero at around $y_1 = 0.8 \times 10^{-15}m$, so we will define our third equation as $F_{EM} = 0$ at $y_1 = 0.8 \times 10^{-15}m$.

$$0 = \frac{3}{2}\frac{\mu}{4\pi}C^2\frac{D^2}{y_1^4}(Q - q + (Q + q)x)^2 - \frac{4k}{y_1^2}((Q - q)^2 + 2Qq\frac{(Dx)^2}{y_1^2+(Dx)^2})/ :k$$

And $\frac{\mu C^2}{4\pi k} = \frac{10^{-7}\frac{N}{A^2}\times(3\times10^8\frac{m}{s})^2}{9\times10^9\frac{Nm^2}{C^2}} = 1$ (if we take that $A = \frac{C}{s}$)

$$0 = \frac{3}{2}\frac{D^2}{y_1^4}(Q - q + (Q + q)x)^2 - \frac{4}{y_1^2}((Q - q)^2 + 2Qq\frac{(Dx)^2}{y_1^2+(Dx)^2}) \qquad /\times y_1^2$$

$$0 = \frac{3}{2}\frac{D^2}{y_1^2}(Q - q + (Q + q)x)^2 - 4((Q - q)^2 + 2Qq\frac{(Dx)^2}{y_1^2 + (Dx)^2})$$

For zero force at $y_1 = 0.8 \times 10^{-15}m$

$$0 = \frac{3}{2}\frac{D^2}{6.4 \times 10^{-31}m^2}(Q - q + (Q + q)x)^2 - 4((Q - q)^2 + 2Qq\frac{(Dx)^2}{6.4 \times 10^{-31}m^2 + (Dx)^2})$$

Or, Third equation:

$$D^2(Q - q + (Q + q)x)^2 - 1.707 \times 10^{-30}\left((Q - q)^2 + 2Qq\frac{(Dx)^2}{6.4 \times 10^{-31}m^2 + (Dx)^2}\right)$$

$$= 0$$

50

I will derive fourth equation from conditions for a minimum of function: We know that the nuclear force has extreme point somewhere around $y_2 = 10^{-15}m$, in other words $\frac{\partial F_{EM}}{\partial y} = 0$ at $y_2 = 10^{-15}m$

$$\frac{\partial}{\partial y}\left(\frac{3}{2}\frac{\mu}{4\pi}C^2\frac{D^2}{y^4}(Q - q + (Q + q)x)^2 - 4k\left(\frac{Q^2}{y^2} + \frac{q^2}{y^2} - \frac{2Qq}{y^2 + (Dx)^2}\right)\right) = 0$$

And for $\frac{\mu c^2}{4\pi k} = 1$, we get:

$$\frac{3}{2}\frac{D^2}{y^5}(-4)(Q - q + (Q + q)x)^2 - 4\frac{(-2)}{y^3}\left(Q^2 + q^2 - 2Qq\frac{y^4}{(y^2+(Dx)^2)^2}\right) = 0 \quad / \quad \times\frac{y^5}{(-4)}$$

For $y = y_2 = 10^{-15}m$ and after some reduction, we get:

The fourth equation:

$$D^2(Q - q + (Q + q)x)^2 - 1.333 \times 10^{-30}\left((Q - q)^2 + 2Qq\frac{2 \times 10^{-30}(Dx)^2 + (Dx)^4}{(10^{-30}m^2 + (Dx)^2)^2}\right) = 0$$

If we simply place these four equations into a computer program, we will get one real but not physically justified solution. With better computer and program under restrictions (Q>q, Q>0 and q>0, and so on) we can find solution that is more physically justifiable. With my old computer and calculating program, I have found that there are regions of physically acceptable solutions for $10^{-16}C \le Q < 10^{-1}C$ (other magnitudes are easily calculable from this one). Let us assess one real solution that is most in sync with mathematics and the one that has real physical meaning.

We will subtract equations 3 and 4 and incorporate equation 1 into situation.

After some work we get first equation for iteration:

$$1.9916 \times 10^{-40} = 2Q(Q - 2.667 \times 10^{-20})d^2\left(\frac{2 \times 10^{-30} + d^2}{(10^{-30}m^2 + d^2)^2} - 1.28\frac{1}{6.4 \times 10^{-31}m^2 + d^2}\right)$$

Where we defined d as $d = Dx$

The second iteration equation is actually our second equation, combined with the first one:

51

$$\frac{Q}{Q - 2.667 \times 10^{-20}} = \left(\frac{(D+d)^2 + 2.5 \times 10^{-29}m^2}{(D-d)^2 + 2.5 \times 10^{-29}m^2}\right)^{\frac{3}{2}}$$

The third iteration equation is the one for zero of the nuclear force; we need to choose physically meaningful solution that is within a frame of our equation.

$$F_{EM} = \frac{3}{2}\frac{\mu}{4\pi}C^2\frac{D^2}{y^4}(Q - q + (Q+q)x)^2 - \frac{4k}{y^2}((Q-q)^2 + 2Qq\frac{(Dx)^2}{y^2+(Dx)^2})$$

This equation should have a zero at around $y_1 = 0.8 \times 10^{-15}m$

After a gazillion iterations I assessed the following set of unknown magnitudes Q, q, x and D only by fine tuning original two iteration equations as:

The first equation as:

```
solve[
  {1.9916 * 10⁻⁴⁰ == 2 * 0.357 * 10⁻¹⁸ * (0.357 * 10⁻¹⁸ - 2.667 * 10⁻²⁰) * d² *
    ( 2 * 10⁻³⁰ + d²
      ─────────────  - 1.28 *        1
      (10⁻³⁰ + d²)²            ──────────────── )}, {d}]
                                6.4 * 10⁻³¹ + d²
```

Which for $Q = 0.357 \times 10^{-18}C$ gave me one of the solutions for d as:

$d = 4.11894 \times 10^{-16}m$

By placing this d and $Q = 0.357 \times 10^{-18}C$ into second iteration equation, with fine toning it, I received D as: $D = 4.19182 \times 10^{-16}m$

```
solve[{( 0.357 * 10⁻¹⁸
         ─────────────────────────  ) == ( (D + 4.11894 × 10⁻¹⁶)² + 0.13 * 10⁻²⁸ )^(3/2)
         0.357 * 10⁻¹⁸ - 2.667 * 10⁻²⁰     ──────────────────────────────────────       }, {D}]
                                           (D - 4.11894 × 10⁻¹⁶)² + 0.13 * 10⁻²⁸
```

Note two things: D have to be larger than d, in order to have any real physical meaning hence fine-tune equation by placing $(2y_0)^2 = 0.13 \times 10^{-28}m$ into previous equation, which gave us distance at y axis at which the electric field is zero as $y_0 \approx 1.8 \times 10^{-15}m$ appose to presumed $y_0 \approx 2.5fm \approx 2.5 \times 10^{-15}m$ from page 42.

With all of that, the nuclear force equation looks as:

$$F = 1.16916 * 10^{-57} * \frac{1}{y^4} - \frac{1}{y^2} * \left(2.56064 * 10^{-29} + \frac{1.44052 * 10^{-57}}{y^2 + 1.69657 * 10^{-31}}\right)$$

This force has two symmetrical real zeros at:

$y_1 = 8.22738 \times 10^{-16} m$ And $y_{1,2} = -8.22738 \times 10^{-16} m$

This is close enough to our original assumption of $y_1 = 0.8 \times 10^{-15} m = 8 \times 10^{-16} m$.

Now let us calculate the unknown magnitudes:

$Q = 0.357 \times 10^{-18} C = 3.57 \times 10^{-19} C$

$q = 0.357 \times 10^{-18} C - 2.667 \times 10^{-20} C = 3.3033 \times 10^{-19} C$

$$D = 4.19182 \times 10^{-16} m$$

$$x = \frac{d}{D} = \frac{Dx}{D} \approx \frac{4.11894 \times 10^{-16} m}{4.19182 \times 10^{-16} m} \approx 0.9826$$

Note that the separation between d and anti-d quark is just around two times larger than D (see the picture above). This means that in d quark positive charges rotate almost at center but the separation between the negative charges is: (see the picture above)

$$D' \approx 2\frac{D}{2} + 2\frac{Dx}{2} = D + Dx = D(1 + x) \approx 4.19 \times 10^{-16} m(1 + 0.9826) \approx 8.307 \times 10^{-16} m$$

If we compare this size with the distance between d and anti-d quark $y_1 = 8.22738 \times 10^{-16} m$

Then we see that we have small problem because the magnetic part of nuclear force equation is derived from let's say $D' \ll y_1$. Problem is deflected if we understand that the magnetic part of the force is derived by the description of four electric current interactions:

The first current interaction is indeed between the rotating charges Q and –Q, and our equation does not describe this interaction very well (this kind of interaction is similar to: see next chapter (d quark stability)).

But other three current interactions are described very well with this equation, i.e. the interaction between the currents:

1) Originated by rotation of +q charges within a d quark and –q charges within an anti-d quark.

2) Originated by rotation of +q charges within a d quark and +Q charges within an anti-d quark.

3) Originated by rotation of -Q charges within a d quark and -q charges within an anti-d quark.

A separation between positive charges within d quark (and between negative charges within anti-d quark) is:

$$D" \approx \frac{D}{2} - 2\frac{Dx}{2} + \frac{D}{2} - 2\frac{Dx}{2} = D - Dx = D(1 - x) \approx 4.19 \times 10^{-16}m(1 - 0.9826)$$
$$\approx 7.29 \times 10^{-18}m$$

Which is indeed $<<y_1 = 8.22738 \times 10^{-16}m$, so the overall equation is acceptable because for at least one of the interacting current, in every interaction 1,2,3 holds $D" \ll y_1$.

In addition to that d quark in the proton is in reality, in my speculative assessment, at least 10 larger in size and mass (charges are same-they wary very little) of anti-d quark within up quark within a neutron. Fill free to incorporate this size changes into the above equations, it will lead us to a better assessment for our unknown magnitudes for sure (I can't do this at this point, I already have a feeling that I overburdened the book with the story about the quark, and there is more about it that I have to show.)

Let us graph our nuclear force (color force is very similar), and calculate the bond energy between d and anti-d quark in our equation.

Assessed force:

$$F = 1.16916 * 10^{-57} * \frac{1}{y^4} - \frac{1}{y^2} * \left(2.56064 * 10^{-29} + \frac{1.44052 * 10^{-57}}{y^2 + 1.69657 * 10^{-31}}\right)$$

(It has a zero at $y_1 = 8.22738 \times 10^{-16}m$) Let us graph this equation near this point:

$$\text{Plot}\left[1.16916*10^{-57}*\frac{1}{y^4}-\frac{1}{y^2}*\left(2.56064*10^{-29}+\frac{1.44052*10^{-57}}{y^2+1.69657*10^{-31}}\right),\{y,6*10^{-16},6*10^{-15}\},\right.$$
$$\text{AxesOrigin}\rightarrow\{4*10^{-16},0\}\right]$$

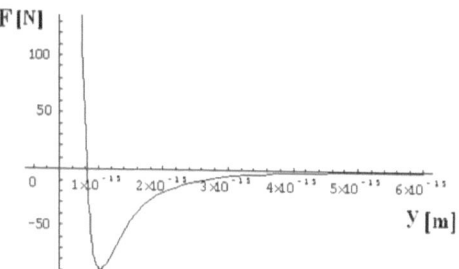

Force equation has its zero at: $y_1 = 8.22738 \times 10^{-16}m$

Its extreme value for $y_2 = 1.02849 \times 10^{-15}m$ is $F = -88.7806$ N

Bond energy = work required for the separation of the d and anti-d quark in this configuration is:

$$W = -\int_{1.02849\times10^{-15}}^{1}\left(1.16916*10^{-57}*\frac{1}{y^4}-\frac{1}{y^2}*\left(2.56064*10^{-29}+\frac{1.44052*10^{-57}}{y^2+1.69657*10^{-31}}\right)\right)dy$$

$W = 6.99 \times 10^{-14}J$

Which correspond to $W \approx 6.99 \times 10^{-14} \times 6.25 \times 10^{18} \approx 440000\ eV \approx 0.44MeV$

If we compare to deuteron bond energy of 2.2MeV [37], we see that our nuclear force is relatively close to the one we anticipated, but its bond energy is roughly $\frac{2.2}{0.44} = 5$ times smaller.

If we forget about iteration equations for a moment and bluntly place three times larger Q;

$Q = 3 \times 0.357 \times 10^{-18}C = 1.071 \times 10^{-18}C$

$q = 1.071 \times 10^{-18}C - 2.667 \times 10^{-20}C = 1.04433 \times 10^{-18}C$

Our force equation:

$$F_{EM} = \frac{3}{2}\frac{\mu}{4\pi}C^2\frac{D^2}{y^4}(Q - q + (Q+q)x)^2 - \frac{4k}{y^2}(Q^2 + q^2 - 2Qq\frac{y^2}{y^2+(Dx)^2})$$

Will become:

$$F = 1.05132 * 10^{-56} * \frac{1}{y^4} - \frac{1}{y^2} * \left(2.56064 * 10^{-29} + \frac{1.36625 * 10^{-56}}{y^2 + 1.69657 * 10^{-31}}\right)$$

(It has a zero at $y_1 = 7.50337 \times 10^{-16}m$) Let us graph this equation near this point:

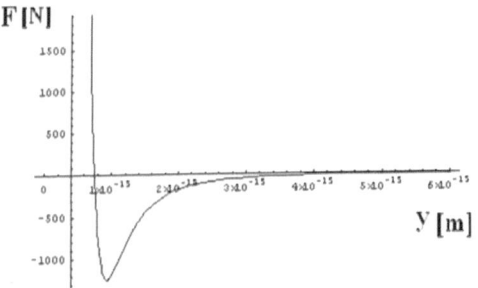

```
Plot[1.05132 * 10^{-56} * 1/y^4 - 1/y^2 * (2.56064 * 10^{-29} + 1.36625 * 10^{-56}/(y^2 + 1.69657 * 10^{-31})), {y, 6 * 10^{-16}, 6 * 10^{-15}},
AxesOrigin → {4 * 10^{-16}, 0}]
```

Force equation has its zero at: $y_1 = 7.50337 \times 10^{-16}m$

Its extreme value for $y_2 = 9.33098 \times 10^{-16}m$ is $F = -1244.62$ N

Bond energy = work required for separation of the d and anti-d quark in this configuration is:

$$W = -\int_{9.33098 \times 10^{-16}}^{1} \left(1.05132 * 10^{-56} * \frac{1}{y^4} - \frac{1}{y^2} * \left(2.56064 * 10^{-29} + \frac{1.36625 * 10^{-56}}{y^2 + 1.69657 * 10^{-31}}\right)\right) dy$$

$W = 7.43 \times 10^{-13}$ J

Which corresponds to W$\approx 7.43 \times 10^{-13} \times 6.25 \times 10^{18} \approx 4600000 \ eV \approx 4.6 \ MeV$

If we compare to deuteron bond energy of 2.2MeV [37], we see that this energy is roughly $\frac{4.6}{2.2} = 2$ times larger than the one that holds proton and neutron together.

Conclusions, guidelines for real anatomy of d quark, anti-d quark, nuclear and color force.

Story behind the listed phenomenon is, of course, much more complex. But first, allow me to repeat myself, when I say:"rotating charges -Q and +q within a d quark" and similar, I do not think about some real pebble like particle that goes around in circles with the speed of light. I say: that is what we observe, measure and feel; not what is real. In reality, the electric charges, if any, oscillate perpendicularly to the motion of quark around itself.

Better results are possible with very similar, but better set of the initial equations and constructions. Firstly, there are five particles within a proton (one d quark and four anti-d quarks; two anti-d quarks form one up quark) and there are four particles within a neutron (two down quarks and two anti-down quarks; two anti-d quarks form one up quark) all of this particles interact electromagnetically with each other, not just two neighbors (let us mention here, electron as a composite particle that is made of three down quarks). Secondly, the d quark in a proton and neutron is larger in size and mass than an anti-d quark within an up quark. Thirdly, our d and anti-d quarks in proton and neutron most likely have two different points in y axis at which electric field is equal to a zero (other zero E points around our d and anti-d quark are probable):

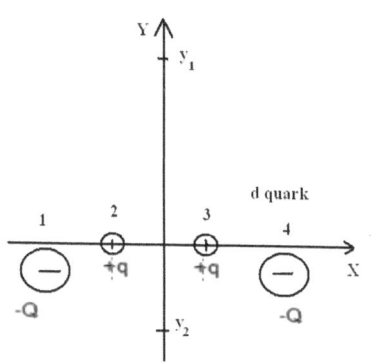

These two different points are possible if negative and positive charges are displaced (in a direction of y axis or in an opposite direction) relative to one another. (At y_1 and y_2 electrical field is zero. $|y_1|>|y_2|$)

In my opinion, this displacement is the reason behind and cause of one of the major difference between the color and the nuclear force.

The bond that holds two anti-d quarks within an up quark and the bond that holds three d quarks within one electron are similar and strictly electromagnetic in its nature; and even though I have some idea about it, I will not address these problems in this book.

The stability of the d and anti-d quark

In all this and future stories I will consider photon and d quark as an unbreakable particle, in a sense that you can destroy or create it, under known conditions, but you can't split them in half. Uncertainty principle [38] holds, but it is something completely different and bear in mind that the Plank's constant is in fact variable).

Let us return to our planar d quark, and use data from the previous calculations.

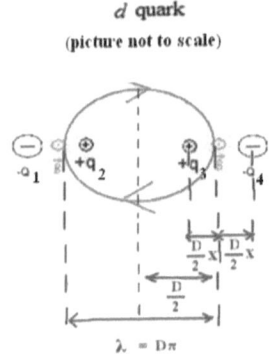

d quark
(picture not to scale)

$Q \approx 3.57 \times 10^{-19} C$

$q \approx 3.3033 \times 10^{-19} C$

$D \approx 4.19 \times 10^{-16} m$

$$x = \frac{d}{D} = \frac{Dx}{D} \approx \frac{4.12 \times 10^{-16} m}{4.19 \times 10^{-16} m} \approx 0.98$$

But first, let us asses the electromagnetic force that is exerted on two electric charges Q_A and Q_B that rotate around each others at distance R (again, these are not real circularly moving electric charges, but a construction similar to our d quark).

Approximation of the Lorentz force [39] exerted on charge Q_A by moving charge Q_B:

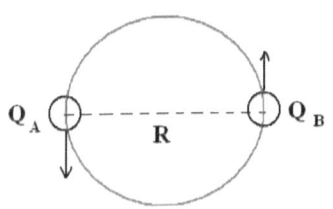

$$F_{AB} = Q_A E_B + Q_A V_A B_B = Q_A k \frac{Q_B}{R^2} + Q_A V_A \frac{\mu I_B}{2\pi R} \quad [40]$$

$$F_{AB} = k \frac{Q_B}{R^2} + Q_A V_A \frac{\mu}{2\pi R} \frac{Q_B V_B}{R\pi}$$

For $V_A = V_B = C$, and because: $\frac{\mu c^2}{4\pi} \approx k$

$$F_{AB} = k \frac{Q_B Q_A}{R^2} \left(1 + \frac{2}{\pi}\right)$$

If our two charges are of the same sign, they will be electrically repulsive and magnetically repulsive (because they will move in the opposite directions).

58

If our two charges are of the opposite sign they will be electrically attractive and magnetically attractive (because they will move in opposite directions).

With all that in mind, the overall magnitude of the electromagnetic force exerted on the charges q_3 and Q_4 ("right" side of the d quark, see the picture on page 52) by moving charges q_2 and Q_1 ("left" side of the d quark, see the picture on page 52) is:

$$F_{EM} = \left[k \frac{q_2 q_3}{\left(2\left(\frac{D}{2} - \frac{D}{2}x\right)\right)^2} - k \frac{Q_1 q_3}{\left(\frac{D}{2} + \frac{D}{2}x + \frac{D}{2} - \frac{D}{2}x\right)^2} + k \frac{Q_1 Q_4}{\left(\frac{D}{2} + \frac{D}{2}x + \frac{D}{2} + \frac{D}{2}x\right)^2} - k \frac{Q_4 q_2}{\left(\frac{D}{2} - \frac{D}{2}x + \frac{D}{2} + \frac{D}{2}x\right)^2} \right] \left(1 + \frac{2}{\pi}\right)$$

In our d quark $q_2 = q_3 = q$ and $Q_1 = Q_4 = Q$

$$F_{EM} = \frac{k\left(1 + \frac{2}{\pi}\right)}{D^2} \left[\frac{q^2}{(1-x)^2} + \frac{q^2}{(1+x)^2} - 2qQ\right]$$

If we place our listed magnitudes within an equation, we will get repulsive EM force as:

$F_{EM} = 2.29 \times 10^7 N$ (Note that for same Q and q our nuclear force at its most attractive position has value of 88.78 N; in reality, most likely, both are larger.)

So, our gravitational force F_G has to overcome repulsive EM force and needs to provide enough pool to keep our d quark in circular motion around itself.

$Force_{centripetal} = F_G - F_{EM}$ Or

$Force_{centripetal} + F_{EM} = F_G$

Let us calculate the left side of the equation, because we have all needed magnitudes.

$$Force_{centripetal} + F_{EM} = \frac{\frac{m_d}{2} c^2}{\frac{D}{2}} + F_{EM}$$

$$\approx \frac{\frac{1.78 \times 10^{-30} kg}{2} (3 \times 10^8 \frac{m}{s})^2}{\frac{4.19 \times 10^{-16} m}{2}} + 2.29 \times 10^7 N \approx 382.34 N + 2.29 \times 10^7 N \approx 2.29 \times 10^7 N$$

In other words, our gravitational force has to be:

$$F_G = 2.29 \times 10^7 N$$

This is very strong gravitational force for such small particle, but none the less, we know that it has to be much stronger than the nuclear force; otherwise the nuclear force would untangle the d quark. A singular possible solution for this situation is that only a small part of the mass of the d quark is caught in a very strong gravitational interaction within a d quark. Let us asses the size R' and the mass m involved in this interaction.

$$F_G = G \frac{\frac{m\prime}{2}\frac{m\prime}{2}}{R\prime^2} \frac{1}{1 - \frac{2Gm\prime}{R\prime C}}$$

Our force is $2.29 \times 10^7 N$, and our mass $\frac{m\prime}{2}$ has to be smaller than a half of the mass of d quark $m_d = 1.78 \times 10^{-30} kg$

So we have only one iteration equation in which we can place mass smaller than $\frac{1.78 \times 10^{-30} kg}{2}$ and see is there any R'<D for which we have a solution:

$$\text{Solve} \left[2.29 * 10^7 = 6.67 * 10^{-11} * \frac{\left(\frac{m\prime}{2}\right)^2}{R\prime^2} * \frac{1}{1 - \frac{2*6.67*10^{-11}*m\prime}{R*(3*10^8)^2}} , R\prime \right]$$

If we place $m' = 1.78 \times 10^{-30} kg$ we get $R' = 1.52 \times 10^{-39} m$, this would mean that the complete mass of the d quark is involved in strong gravitational interaction, which is not realistic (because, if we place D=4.19 × 10⁻¹⁶ m instead of R' within our gravitational equation, we get force of $3.14 \times 10^{-40} N$, which is close to nothing.)

Much more realistic scenario is that the only very small part of the mass (size) is involved in a strong gravitational interaction.

Let's say that $m' = \frac{1.78 \times 10^{-30} kg}{1000}$, what the assessment is for R' in that situation.

$$\text{Solve} \left[2.29 * 10^7 == 6.67 * 10^{-11} * \frac{\left(\frac{1.78*10^{-30}}{2*1000}\right)^2}{R\prime^2} * \frac{1}{1 - \frac{2*6.67*10^{-11}*\frac{1.78*10^{-30}}{1000}}{R\prime*(3*10^8)^2}} , R\prime \right]$$

$R' = 1.52 \times 10^{-42} m$

This is much more realistic scenario, and we say: within a region deep inside the d quark (R'Within aD, i.e.$1.52 \times 10^{-42}m$ within$4.19 \times 10^{-16}m$).

Thousands of parts of the d quark are in such dense condition that the attractive gravitational force becomes dominant on all others.

A more precise assessments for these parts R' and m' are possible, if we take into account different reduction of electric charges Q and q near very strong gravitational field.

Magnetic dipole moment and angular momentum of a d quark

Magnetic dipole moment M

Magnetic dipole moment [26] M of the current loop, I –current, A-area encircled by current: $M = IA$

$$M = I_{Q'}A_{Q'} - I_{q'}A_{q'}$$

$Q' = 2Q$-a magnitude of the overall negative electric charge within the d quark.

$q' = 2q$-a magnitude of the overall positive electric charge within the d quark.

$I_{Q'}$-the current caused by "circular" motion of the negative electric charge.

$I_{q'}$- The current caused by "circular" motion of the positive electric charge, in the same direction as the positive electric charge. Magnetic field of these two currents will subtract, because currents are motions of different charges in same direction.

$$M = \frac{Q'}{T_{Q'}}A_{Q'} - \frac{q'}{T_{q'}}A_{q'} = \frac{Q'}{2\pi R_{Q'}}\pi R_{Q'}^2 - \frac{q'}{2\pi R_{q'}}\pi R_{q'}^2 = \frac{Q'CR_{Q'}}{2} - \frac{q'CR_{q'}}{2}$$

$$= \left(\frac{2Q(\frac{D}{2}+\frac{D}{2}x)}{2} - \frac{2q(\frac{D}{2}-\frac{D}{2}x)}{2}\right)C$$

$$M = \left(\frac{2Q\left(\frac{D}{2}+\frac{D}{2}x\right)}{2} - \frac{2q\left(\frac{D}{2}-\frac{D}{2}x\right)}{2}\right)C = \frac{D}{2}C(Q + Qx - q + qx) = \frac{D}{2}C(Q - q + x(Q + q))$$

$$M = \frac{4.19 \times 10^{-16}m}{2}(3 \times 10^8 \frac{m}{s})(2.667 \times 10^{-20}C + 0.98(3.57 \times 10^{-19}C + 3.3033 \times 10^{-19}C))$$

$M = 4.40 \times 10^{-26} \text{Am}^2$ or $M = 4.40 \times 10^{-26} \frac{J}{T}$

Compare with Bohr's magneton: $M_B = 9.27 \times 10^{-24} \frac{J}{T}$

Angular momentum L

L is a magnitude of angular momentum of our d quark (angular momentum of photon-like particle that rotates around circle of radius D/2; for photon momentum see "Compton's Equation" [28]).

$$L = \frac{D}{2} m_d C = \frac{4.19 \times 10^{-16} m}{2} 1.78 \times 10^{-30} kg \times 3 \times 10^8 \frac{m}{s} \approx 1.12 \times 10^{-37} \text{Js}$$

Compared with value of the Plank's constant (in our laboratory): $h = 6.626 \times 10^{-34} \text{Js}$

The relation between a magnetic dipole moment and an angular momentum for our d quark:

$$\frac{M}{L} = \frac{4.40 \times 10^{-26} \frac{J}{T}}{1.12 \times 10^{-37} \text{Js}} \approx 3.93 \times 10^{11} \frac{1}{Ts}$$

Comparing this value with relation between Bohr's magnet on and the value of the Plank's constant (in our laboratory):

$$\frac{M_B}{h} = \frac{9.27 \times 10^{-24} \frac{J}{T}}{6.626 \times 10^{-34} \text{Js}} \approx 1.40 \times 10^{10} \frac{1}{Ts}$$

The elemental particles (electron, proton and neutron) owe their respective intrinsic properties [41] to above calculated magnetic dipole moment and angular momentum of our d (and anti-d quark); because electrons, neutrons and protons are particles composed out of d and anti-d quarks (an electron is made of three d quarks etc). My numerical values can be a little off but the solution is certainly in this specific direction.

Relativity

The Einstein's special and general theory of relativity [42], were excellent and correct at its time, but in light of these previous speculations I have to say that:

1. Equations within a special theory of relativity have great degree of accuracy, and as you can see those equations are a corner stone for every assumption in this book, but Einstein's interpretation of equations in special theory of relativity is slightly of course. That course led Him to disregard a variability of Planck's "constant" within gravitational field and other misconceptions.

2. General theory of relativity is synchronized with special theory of relativity as much as possible, that means further away in wrong direction. General relativity:

a) Ignores the rest mass change in the gravitational field.

b) Ignores the energy conservation for a free falling photon.

c) Does not recognize the domain in which a gravitational force becomes repulsive etc.

I do not have the ambition to correct both theories, but next example should help others to do just that.

You stand at the surface of the Earth (point 1) and throw the apple straight up, the apple will fly up to the point 2, there it will stop and start to move back to you. The smallest rest mass of the apple in flight is at point 1, and the largest rest mass of the apple is at the point 2. Thought the flight the rest mass increases from point 1 to point 2, and decreases back to its original value as it falls freely from point 2 to point 1. Contrary to that, its speed (relative to point1) decreases from its larger value at point 1 to zero at point 2, and there it will stop and start to move back to you achieving its previous largest value when it arrives at point 1.

Through the flight energy is conserved and through the flight gravitational interaction of the apple and the Earth is as if apple has constant mass (energy is what creates and feels gravitational field, invariant mass does not change in flight) equal to the rest mass at the point 2.

Or, go back to the previous calculations to see that our work against the gravitational field, in hoisting object from point 1 to point 2 will not vanish, but will

63

result in the increase of the rest mass at point 2; and that difference in energy (left side of the equation below), will be equal to the kinetic energy at point 1 (the right side of the equation below) after object falls freely from point 2 to point 1:

$$m_2c^2 - m_1c^2 = \frac{m_1c^2}{\sqrt{1 - \frac{v_1^2}{c^2}}} - m_1c^2$$

Or in short:

$$m_2c^2 = \frac{m_1c^2}{\sqrt{1 - \frac{v_1^2}{c^2}}}$$

Where m_2- the rest mass at point 2, m_1- the rest mass at point 1, v_1- the speed of the object at point 1 relative to point 1 (or relative to point 2 does not matter, both points are stationary relative to one another).

Guidelines for understanding

First and foremost, the origin of every other assumption is a change of the rest mass in the gravitational field; the other is the energy conservation for the photon in the gravitational field, and the third pillar is my personal belief in quark-ed structure of an electron.

The acknowledgments

Love and gratitude to my Kamerad Biljana for companionship in a lifelong chase for a pot at the end of a rainbow.

I am grateful to certain Chicago physics professor for his guidelines and kindness in a sea of indifference and resentment.

I am grateful to Dr. Nave from Georgia State University for his extensive cross section of physics at http://hyperphysics.phy-astr.gsu.edu/hbase/hframe.html.

Also, I wish to thank Eddie A. Maalouf, LinkedIn "Theoretical Physics [43]" the group owner, for providing stimulating environment for physicists and physics enthusiasts to communicate their ideas. Mister Eddie is someone who performs impossible tasks on a daily basis for free. Imagine handling 46 000 of mostly higher educated individuals; of whom the majority is socially challenged (including me) but are driven to communicate about their things passionately.

In the next book

I will try to offer new perspective to some of the quantum world phenomena, through basic idea of space-particle relation.

Every point of space around us is capable to produce multitude of particles [47]. That kind of space must experience some kind of change if we place a particle in it.

Or, in real particle soundings I expect certain antiparticle-particle deformation of a space; this is in direct link to: Inertia, speed of light as a maximum limit, de Broglie waves [44], uncertainty principle [45], tunneling effect [46] etc.

Simplified visualization

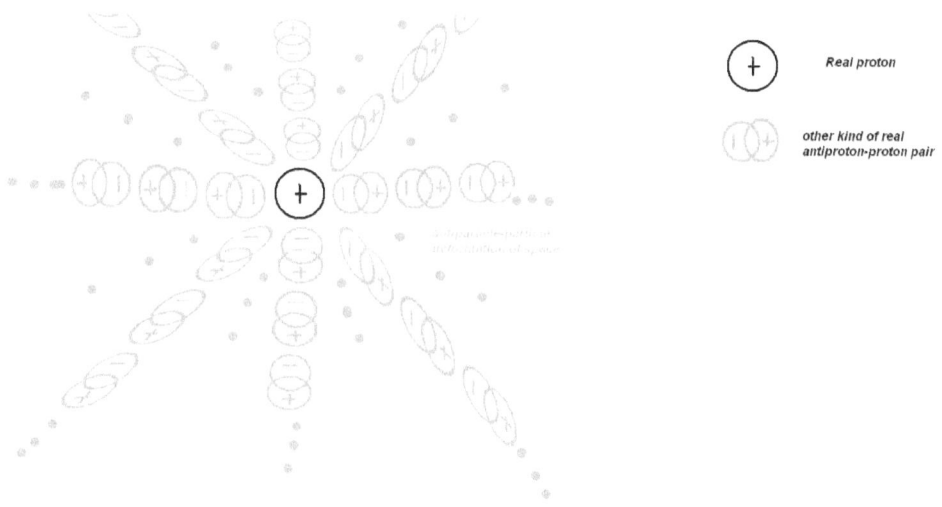

Real proton

other kind of real antiproton-proton pair

References

[1] M. Planck, The Theory of Heat Radiation, Dover Publications (2011)

[2] A. Einstein, Photoelectric effect, Über einen die Erzeugung und Verwandlung des Lichtes betreffenden heuristischen Gesichtspunkt, Annalen der Physik (1905)

[3] D.Bodanis, E=mc^2: A Biography of the World's Most Famous Equation, Bloomsbury Publishing (2009).

A.Einstein , "Über die vom Relativitätsprinzip geforderte Trägheit der Energie", Annalen der Physik (1907)

[4] R. V. Pound and G. A. Rebka, Jr., Gravitational Red-Shift in Nuclear Resonance, Physical review letters (1959)

[5] O.R. Jovanovic, Correct Interpretation for "Harvard Tower Experiment" or Law of Energy Conservation for "Free Fall" Photon, Applied & Computational Mathematics (2015)

[6] M. Salaris and S. Cassisi, Evolution of Stars and Stellar Populations (Wiley, 2005).

[7] C.I. Calle, Einstein For Dummies, Wiley Publishing, Inc. (2006)

[8] Arthur S. Eddington, The Internal Constitution of the Stars (Cambridge Science Classics), ISBN-13: 978-0521337083

[9] R. A. Alpher, H. Bethe, and G. Gamow The Origin of Chemical Elements; Phys. Rev. 73, 803 – Published 1 April 1948;

[10] Physical constants; https://physics.nist.gov/cuu/pdf/all_2002.pdf

[11] Elastic Potential Energy; http://hyperphysics.phy-astr.gsu.edu/hbase/pespr.html

[12] Neutron Star Masses, https://www.lsw.uni-heidelberg.de/users/mcamenzi/NS_Mass.html

[13] Sun Fact Sheet, https://nssdc.gsfc.nasa.gov/planetary/factsheet/sunfact.html

[14] "The relativistic kinetic energy expression can be written as", http://hyperphysics.phy-astr.gsu.edu/hbase/Relativ/releng.html

A.Einstein , "Über die vom Relativitätsprinzip geforderte Trägheit der Energie", Annalen der Physik (1907)

[15] Weisstein, Eric W. "Maclaurin Series." From MathWorld--A Wolfram Web Resource. http://mathworld.wolfram.com/MaclaurinSeries.html

Beyer, W. H. (Ed.). CRC Standard Mathematical Tables, 28th ed. Boca Raton (1987)

[16] CM Claudel, KS Virbhadra, The geometry of photon surfaces, GFR Ellis - Journal of Mathematical Physics (2001)

[17] J.G. Williamson, M.B. van der Mark, Is the electron a photon with toroidal topology?, Annales de la Fondation Louis de Broglie, Volume 22, no.2, 133 (1997)

R. Gauthier, Electrons are spin 1/2 charged photons generating the de Broglie wavelength, 10.13140/RG.2.1.3316.6567 (2015)

[18] A. Das, T. Ferbel, Introduction to Nuclear and Particle Physics, World Scientific (2003)

[19] L. Sodickson; W. Bowman; J. Stephenson; R. Weinstein, Single-Quantum Annihilation of Positrons, Physical Review (1970)

[20] W. Vassen, The proton radius revisited, Science (2017)

[21] A. Cho, Mass of the Common Quark Finally Nailed Down, Science (2010)

[22] "Why is the proton so much more massive than the electron, yet holds the same charge?", Frank Heile, https://www.quora.com/Why-is-the-proton-so-much-more-massive-than-the-electron-yet-holds-the-same-charge/answer/Frank-Heile

[23] Robert Millikan, University of Chicago, Chicago, IL, https://www.aps.org/programs/outreach/history/historicsites/millikan.cfm

[24] S.I.Djenize, Osnovi atomske, kvantne i molekulske fizike, Nauka (1995)

[25] W.Gerlach, O. Stern, Der experimentelle Nachweis der Richtungsquantelung im Magnetfeld, Zeitschrift für Physik (1922).

S.I.Djenize, Osnovi atomske, kvantne i molekulske fizike, Nauka (1995) [26] D. Griffiths, Introduction to Electrodynamics, Prentice Hall (1999)

[27]J.C. Maxwell, A Dynamical Theory of the Electromagnetic Field, Philosophical Transactions of the Royal Society of London (1865)

E. M. Purcell, D. J. Morin, Harvard University, Electricity and Magnetism, Cambridge University Press, (2013)

[28] R.S. Shankland,The Compton Effect with Gamma-Rays, Physical Review Journals, (1937)

S.I.Djenize, Osnovi atomske, kvantne i molekulske fizike, Nauka (1995)

[29] J.C. Hafele, R.E. Keating,Around-the-World Atomic Clocks: Predicted Relativistic Time Gains, Science (1972)

[30] Linear Electric Quadrupolc, http://hypcrphysics.phy-astr.gsu.edu/hbase/electric/elequad.html

[31] proton rms charge radius, https://physics.nist.gov/cgi-bin/cuu/Value?rp

[32] nuclear force, http://www.cyberphysics.co.uk/topics/particle/strong.html

[33] Linear electric quadrupole, http://hyperphysics.phy-astr.gsu.edu/hbase/electric/elequad.html#c1

[34] N. N. Nedeljkovic, Lj. D. Nedeljkovic, Uvod u elektromagnetizam, D.P."Studentski trg" (1992)

[35] A. Zangwill, Modern Electrodynamics, Cambridge University Press (2013) ("Two current rings" -page 399)

[36] Nuclear force, http://www.scholarpedia.org/article/Nuclear_Forces

[37] D. M. Ivanovic, V.M. Vucic, Elektomagnetika i optika, Naucna knjiga (1971)

[38] S.I.Djenize, Osnovi atomske, kvantne i molekulske fizike, Nauka (1995)

[39] D. M. Ivanovic, V.M. Vucic, Elektomagnetika i optika, Naucna knjiga (1971)

[40] D. M. Ivanovic, V.M. Vucic, Elektomagnetika i optika, Naucna knjiga (1971)

[41] S.I.Djenize, Osnovi atomske, kvantne i molekulske fizike, Nauka (1995) http://hyperphysics.phy-astr.gsu.edu/hbase/Nuclear/nmr.html

[42] Einstein, Albert, Ph.D. Professor of Physics at the University of Berlin (1920),

 Relativity: The Special and the General Theory: Popular Exposition; authorised translation by Robert W. Lawson, D.Sc., University of Sheffield

[43]"Theoretical physics", https://www.linkedin.com/groups/3091009/

[44] S.I.Djenize, Osnovi atomske, kvantne i molekulske fizike, Nauka (1995)

[45] S.I.Djenize, Osnovi atomske, kvantne i molekulske fizike, Nauka (1995)

[46] Barrier Penetration, http://hyperphysics.phy-astr.gsu.edu/hbase/quantum/barr.html

[47] Yen-Jie Lee, R. Gunther,W. Bolek, Physics Letters B (2015)

"Particles created from the proton collision stream", http://news.mit.edu/2015/large-hadron-collider-first-results-1014

www.ingramcontent.com/pod-product-compliance
Lightning Source LLC
Chambersburg PA
CBHW021905170526
45157CB00005B/1983